T0235466

Femininity, Mathematics and Science, 1880–1914

Femininity, Mathematics and Science, 1880–1914

Claire G. Jones
Associate Lecturer in History, University of Liverpool

First published 2009 by
PALGRAVE MACMILLAN

Palgrave Macmillan in the UK is an imprint of Macmillan Publishers Limited,
registered in England, company number 785998, of Houndmills, Basingstoke,
Hampshire RG21 6XS.

Palgrave Macmillan in the US is a division of St Martin's Press LLC,
175 Fifth Avenue, New York, NY 10010.

Palgrave Macmillan is the global academic imprint of the above companies
and has companies and representatives throughout the world.

Palgrave® and Macmillan® are registered trademarks in the United States,
the United Kingdom, Europe and other countries.

ISBN 978-1-349-36409-1   ISBN 978-0-230-24665-2 (eBook)
DOI 10.1007/978-0-230-24665-2

This book is printed on paper suitable for recycling and made from fully
managed and sustained forest sources. Logging, pulping and manufacturing
processes are expected to conform to the environmental regulations of the
country of origin.

A catalogue record for this book is available from the British Library.

A catalogue record for this book is available from the Library of Congress.

10   9   8   7   6   5   4   3   2   1
18 17 16 15 14 13 12 11 10 09

For Hywel

Contents

Illustrations, Figures and Tables

Tables

Acknowledgements

This project began life as a doctoral thesis at the University of Liverpool and I am immensely grateful for the guidance given to me there by William Ashworth and Pat Starkey. Their encouragement and support was (and is) unfailing. I would also like to thank Graeme Gooday at the University of Leeds and Eve Rosenhaft at the University of Liverpool; without their generous criticism this book would, without a doubt, be considerably less. However any infelicities or errors are, of course, all mine.

I am in debt to the many archivists without whom this project would not have been possible. Without exception they have given liberally with their time and advice. Special thanks to Kate Perry at Girton College, archivists at The Royal Society of London, and Adrian Allan at the University of Liverpool without whom I would not have discovered one of my leading ladies, Grace Chisholm Young, the mathematician who inspired the entire project.

I know that I would never have completed this book without the loving assistance and calm reassurance of my partner Hywel.

Introduction

It is largely a question of mathematics, – and woman is not
– (*naturally* speaking) – a mathematician. One woman here
and there may occasionally train herself to out-do the sharpest
master of the science, but in the effort she will lose far more
than she gains. And, as in music, so in politics, – woman's
business is to illumine the background – to inspire the work,
and let her light 'shine through' the victorious accomplish-
ment of noble purpose.

Marie Corelli (1907)[1]

The decades around 1900 have been identified by cultural historians as
the years of 'fin de siècle' tension. During a time 'fraught with anxiety
and with an exhilarating sense of possibility',[2] debate raged as the
assertive 'new woman' demanded right of entry to traditionally male
spheres; as supporters of women's higher education challenged
Victorian notions of distinct male and female intellects; and as suffrage reformers accelerated their campaign for the vote. How did this
changing landscape of gender interact with the equally fluid domain of
turn-of-century science and mathematics?

According to the historiography of women and science, the closing
years of the nineteenth century were a time of rapid institutionalisation when women became 'locked out' of the spaces of science such as
laboratories, learned societies and elite university departments. This is
undoubtedly true. But the growth of scientific endeavour did bring
opportunities for participation, even for women. How are we to interpret data that suggests that there were proportionally more women
active in research mathematics at the end of the nineteenth century
than in the 1960s? How do we understand that a woman was nearly

1

elected to a Fellowship of the Royal Society in 1902 yet it was to be over forty years later when this particular milestone was eventually reached?

By exploring the differing cultures of mathematics and science, this book aims to use gender as a window with which to open up new perspectives, asking how, and in what sense, women were able to participate meaningfully within their chosen discipline. Did women always need a male mentor to mediate continuing involvement? Were the increasingly-influential Darwinian sciences, theoretically and in practice, wholly antithetical to femininity and to women? Social historians of science have stressed the role of 'community' and 'trust' in the production of knowledge, but how does gender operate within this model?[3] 'Professionalisation' is often put forward as the reason why women were kept to the periphery, but how did this work? What were the complex mechanisms of inclusion and exclusion around which mathematical and scientific women navigated?

In addressing these questions, a controlling thread is provided by the careers of two women: experimentalist and electrical engineer Hertha Ayrton (1854–1923) and pure mathematician Grace Chisholm Young (1868–1944). These two individuals promise a balanced insight into the operations of gender within their chosen discipline as they hold much more than just their femininity in common. They both studied, in succeeding decades, for the mathematics tripos at Girton College, Cambridge and, unusually for their time, each chose to be a practitioner of her discipline rather than a teacher of it. Furthermore, each woman's politics and philosophies were intimately connected with her science; this was manifested in their intellectual product, the constituency for their work, their views on equality between the sexes and their relationship strategies. That they both married men who pursued the same calling as themselves (and achieved prominence within it) facilitates an analysis of the politics of collaboration between the sexes which illuminates differences between mathematics and practical science at this time.

The pure and the practical

In a study seeking to contrast the differing configurations of mathematics and practical science it is essential – if difficult – to define terms. The word 'practical' has been used in preference to 'applied' as the latter may often refer to work that retains a strong theoretical element. During the nineteenth century all areas of science witnessed a growing distinction between intellectual labour and its practical or industrial application, a distinction which intersected with notions of class and was key to the identities of male and female scientists. In addition, the scope of this study does not extend to the biological sciences as these have their own histories, aesthetics and gender colourings which require separate consideration.

The problems of locating the boundaries between science and technology, science and engineering, or mathematics and 'physical' mathematics, is illustrated by the lack of consensus on the issue and, in one sense, the search is fruitless as the concepts themselves are fluid and subject to historical change. All that can be maintained with any certainty is that the terms imply a certain hierarchy and that, in the decades surrounding 1900 at least, science represented the 'privileged' side of the science-technology relationship.[4] The introduction of mathematics adds another level of complexity to the issue. Where does mathematics end and mathematical physics begin? When does mathematical physics become the same as practical science or experimental science? And when does the latter turn into technology/engineering? These distinctions can become unclear whatever defining criteria are used, whether they be project aims, working practices, theory or experiment, or anything else. Engineering can result in discovery (often taken as the hallmark of 'science') and experimental science can, and often does, result in technology. Around 1900, many members of the scientific community combined roles as researcher with producer of technology, or engineer with experimental investigator; this was in keeping with the unfolding nature of these disciplines and reflects the processes of specialisation. It did not follow, however, that all occupations were regarded as equally valid; tensions and hierarchies, often informed by class, rivalry between new and older traditions, or accusations of commercialism, could and did arise. There is evidence that issues of categorisation were no less confusing to contemporaries intimately connected with the rapidly-developing fields of mathematics and science at turn-of-century. The scientific journal *Nature* alluded to the problem in November 1900 when it reported on the modern 'cleavage in mathematical thought'. *Nature* went on to explain that, at the recent Physical and Mathematical Congresses held simultaneously in Paris, any follower of Maxwell and Kelvin uninterested in the new mathematics of analysis 'must turn to the physical sector for the interest he requires'.[5] It seems clear that contemporaries made distinctions between (and moral judgements on) the pure and practical sciences, even if boundaries remained ill-defined and unstable.

For this study, a working distinction will be made between predominantly laboratory-based investigations, subsumed under the umbrella 'practical science', and mathematics as predominantly desk-based. While not implying any rigid categories, this does reflect crucial differences in material and epistemological practice; it also helps to reveal important historical differences between the two in configuration and self-identity which were central to the participation of women. Mathematics did not professionalise in the same way as the sciences and did not transfer to new spaces of practice such as the laboratory – places where, as this study will show, any hint of femininity came to be seen as incongruous and inappropriate.

Scholarly context

Much of the scholarship on women and mathematics is shaped around biography, although there is a growing body of working offering more theoretical insights.[6] In contrast, social studies of science have proliferated in recent years. There is still debate about what pure mathematics is and how it relates – if it does – to reality. Logician and philosopher Bertrand Russell wrote in 1901 that mathematics was not a natural science but the intellectual creation of man. Another leading mathematician of the nineteenth century, J.J. Sylvester, held a similarly 'intuitionist' view of pure mathematics as 'unfolding the laws of human intelligence'.[7] Implicit in these ideas, and in many histories of mathematics, is the idea that mathematics is absolute knowledge – permanent and unchanging over time. Such an assumption can lead to a neglect of the social and cultural implications of its production, although significant studies have and are being done in this area.[8] If attention is focused on disembodied mathematical ideas at the expense of context, only a limited account of mathematical change can be offered. Such an approach cannot, by definition, address many of the issues raised here, for example the changing relations between gender and mathematics, and science and mathematics, around 1900. One of the assumptions of this study is that mathematics is informed by its social and cultural context (and that, in turn, mathematical ideals and practices spill over to inform the society and politics of their practitioners). Once we examine mathematics from this perspective, gender becomes one of the factors to be addressed in assessing why value is accorded to some results rather than others, or why one area of mathematics is considered more important than another. It is argued here that pure mathematics became less conflicted with contemporary prescriptions of femininity around 1900, partly in response to the rise of professional engineering and science which defined itself via an active, virile masculinity. In this way, the idea that mathematics is always configured as an innately masculine discipline is challenged, acknowledging that this gender 'colouring' is concerned with the culture of mathematics, not the fact that participation may be dominated by one particular sex.

There is a larger body of scholarship on gender and science than on gender and mathematics. Science is made up of many disciplines offering opportunity for case study, plus mathematical women are often subsumed under the umbrella of 'women in science' in biographical anthologies and in important, more analytic, contributions too.[9] In addition, there is a valuable collection of case studies revealing a more rounded picture of women in science by uncovering previously 'invisible' women and following them into the laboratory, or into their marriages and collaborative partnerships.[10]

Acknowledging that the epistemologies and cultures of mathematics and science differed markedly around 1900, here women mathematicians and scientists are considered separately with the aim of illuminating more precisely the experiences of both. At the end of the nineteenth century, mathematics and science were engaged in a struggle for status and influence which resulted in self-definitions and legitimations that relied on 'difference' from the other. Pure mathematics retreated into abstraction and prided itself on being 'uncontaminated' by the real world; in contrast, experimental science and engineering pointed to its utility in the real world as a generator of progress and technology, and as a source of explanation about nature. This book explores how these differing self-definitions affected the position of women and informed the intellectual product of each discipline.

Book structure

Although Grace Chisholm Young and Hertha Ayrton are used as an anchor for the text, the book is not written as any kind of biographical narrative, neither is it based exclusively on their experiences. The two women are referred to predominantly by their first names as this makes it easier to distinguish them from their spouses and, in Hertha's case at least, reflects a desire often articulated to retain a separate identity from her husband.[11] An inclusive approach has been adopted with reference to sources; those used extend beyond institutional records, letters and memoirs, to include literary, journalistic and photographic/pictorial evidence too. There has been much attention paid recently to uncovering women mathematicians and scientists and quantifying their numbers and contributions.[12] While acknowledging a debt to this meticulous research, this study has largely different aims and seeks to foreground the issues of *why?* and *how?* rather than *how many?*

Chapter 1 assesses the significance of women training for the mathematics tripos at Cambridge University. Reflecting this examination's importance as a training ground for many of the greatest figures in the history of British physics, there has emerged a growing body of scholarship on the history of Cambridge mathematics in the nineteenth and early twentieth centuries.[13] During much of this time, the mathematics tripos was *the* prestige degree for men, inextricably bound up with ideals of masculine intellect, physical stamina and middle-class aspiration. Previous studies have assessed the tripos from the vantage point of men; here the focus shifts to women, revealing how Cambridge mathematics adapted to femininity – and how femininity impacted on this elite degree – including differing coaching dependent on sex and the development of gendered notions of success. This chapter also provides an introduction to Hertha and Grace, with some biographical background.

Chapter 2 moves from Cambridge to Germany and the University of Göttingen, an institution whose standing in the mathematical world around 1900 was epitomised in its reputation as 'the Shrine of Pure Thought'. At a time when a struggle was underway over the foundations of mathematics, Göttingen-style abstraction was beginning a transformation of the discipline which would eventually revolutionise mathematics at Cambridge and elsewhere. Using the extensive archive of doctoral candidate Grace, and other sources, this chapter recreates the experiences, motivations and strategies of mathematical women studying at this elite institution in the 1890s and early 1900s. But femininity is just half the story and this chapter is also an analysis of the 'special' masculinity of mathematical men. Even beyond the University, Göttingen was a town built on the reputations of its mathematical heroes and a place where the spirit of romanticism still held sway. Here, the worship of 'intellectual' masculinity provided a dynamic which helped mould the practises and ideals of the School of Mathematics.

By moving from a world of abstraction to the 'hands-on' mathematics and engineering of the Central Institution in South Kensington, Chapter 3 delineates the accelerating opposition between practical or experimental concerns and pure mathematics, exposing the very different interpretations of gender within each discipline. Here, the experiences of Hertha Ayrton as a student of electricity and researcher are recreated, providing an insight into women and technical training at the end of the nineteenth century. As in the preceding chapter, masculinity is a key concern and the 'active masculinity' of the engineers is contrasted to that of the 'pure' mathematicians. The collaborative practises of the Ayrtons, in particular the 'public face' they sought to project, were in marked contrast to the partnership of the Chisholm-Youngs explored in Chapter 4. Within the context of a reassessment of the latter couple, this chapter illustrates how the 'male-mentor' model of women's access and participation in 'male' spheres can obscure a more complex picture. For the Chisholm-Youngs, the 'mentor' was Grace.

Chapter 5 moves to the laboratory, illustrating how the contrasting cultures of the pure and the practical served to define and limit women's involvement in different ways. Hertha's experiences in the laboratory are examined with reference to the growth of an active, heroic culture around the experimental space which gave it a specifically masculine colouring. Chapters 6 and 7 attempt a broader analysis, seeking to place both women within the context of their disciplines and identify patterns of involvement for women as a whole. In Chapter 6, particular attention is given to the London Mathematical Society and publishing opportunities. Historians have long pointed to a connection between masculinity and abstract rationality on the one hand, and femininity and the emotions on the other, and have explained how these dualities have led to the exclusion of women from mathematics. The tenacity and ahistoricism of such connections is questioned here, as a growing affinity between pure mathematics and femininity in the late nineteenth century is uncovered and explored.

Chapter 7 assesses the culture of the Royal Society and places the issue of gender firmly within the context of factional infighting and changes within society at large (suffrage, the science of gendered intellect, women's education). Who should own the soul of science, pure researchers or engineers? Businessmen, technologists or academics? Gentlemen of science or the new breed of paid scientist? Whether or not to argue the women's cause became important to the identity of a male fellow and a powerful tool in his pursuit of influence and reform. Together, Chapters 6 and 7 facilitate an understanding of the contrasting contexts of science and mathematics and the ways in which these offered different opportunities and limitations to women. In conclusion, Chapter 8 brings the threads of analysis together to argue that gender became a pivotal issue for both the pure and the practical, especially in battles to establish contrasting identities and elite status. Gender – femininity and masculinity – is not peripheral to the social history of science and mathematics, it is fundamental.

1

The 'Glamour' of a 'Wrangler': Women and Mathematics at Girton College, Cambridge[1]

Emily Davies, principal founder of Girton College, the Cambridge College of higher education for women, observed in 1868 that 'the best girls' schools are precisely those in which the 'masculine' subjects have been introduced'.[2] The subjects that she was referring to were mathematics, Latin and Greek. When it came to the contentious issue of higher education for women, Davies was convinced that only if women succeeded in subjects held to be prestigious for men would their educational achievements be recognised as equally valid. She rejected any idea of a special system or curriculum for women because, to opponents of women's higher education, 'different' would automatically mean 'inferior'. Davies put her ideas into practice the following year when she opened a residential college for women at Benslow House in Hitchin; these premises soon proved too small and in 1873 a move was made to new purpose-built buildings, some four miles distant from Cambridge, and Girton College was born. Here Davies encouraged her students to study for the most highly-regarded triposes, especially mathematics, a subject that had long been a symbol of masculine success and which was to remain the elite Cambridge degree until the 1890s.

The connections between Cambridge mathematics and notions of superior English masculinity were long-standing and profound. To mid-century and beyond, the middle and upper classes sent their sons to study mathematics at Cambridge in order to train them as gentlemen, not to turn them into mathematicians. Mathematics, as an essential component of a 'liberal education', was believed to train the character and the intellect, producing the fair judgement and unclouded mind necessary for men who were to assume their rightful, elevated place in society and Empire. Unlike leading European centres of mathematics, Cambridge adhered to a proud Newtonian tradition of 'mixed mathematics' which shunned newer, more specialist techniques of abstract analysis in favour of solid, physical and geometrical approaches to problem solving. So important was such training believed to be that, until 1850, it was mandatory even for those taking final honours in classics to have taken the mathematical tripos first.

Accounts left by candidates leave no doubt as to the physical and psychological pressures experienced both in preparing for and sitting this highly-competitive examination. On the day(s) this involved working against time to find solutions to problems which became ever harder as the examination progressed. These problems ranged across a wide range of topics and were presented in the form of open-ended papers sat over a series of consecutive days. Students were 'trained like racehorses' as the tripos 'rivalled the Newmarket races, and the bets on the outcome were just as keen'.[3] As befits such a competitive sprint, students were then individually ranked according to their performance in an order of merit which was announced publicly at Senate House, often before a rumbustious crowd. After 1882, when papers were made available to women on a formal basis, female students were ranked alongside the men, although women had the right to examinations only, not to degrees. To be 'senior wrangler' (or first among the first class) was for the man who achieved it a path to opportunity, often leading to a coveted Cambridge fellowship or high office in another field or profession.[4]

A small collection of scholarship exists on the mathematics tripos at Cambridge University in the second half of the nineteenth century. Special attention has been paid to the development of teaching methods, its relationship to the natural sciences tripos, and the connections between the severe mental and bodily drill involved in preparing for the examination.[5] Little attention has been given to the experiences or significance of women training for the tripos and, for the most part, the history of mathematics, and of mathematicians, has been written as an unthinkingly gendered narrative. Such approaches to the history of mathematics seem to rest on an assumption that the discipline is, by its very nature, intrinsically masculine. Towards the end of the nineteenth century such views were not just assumed tacitly, but argued loudly. Rational, often abstract, eminently cerebral and never emotional, mathematics was a man's subject generally held to be altogether too hard for women. Sciences informed by Darwinian understandings of male and female nature were becoming increasingly influential and these pointed to woman's less evolved brain, decreased capacity for abstraction, and greater subjection to the emotions. Her biology had evolved for reproduction and her brain worked on instinct; put her in an environment designed for men, set her to masculine intellectual work – especially mathematics – and her health, even her feminine appearance, could be at risk.[6] As Sara Burstall, a 'graduate' of the 1881 tripos recalled, ' … to much of Cambridge 'Varsity opinion there was something comic in teaching girls mathematics at all'.[7] Indeed, the familiar cartoons of the time featuring stern, manly and monocle-wearing women, often 'Girton girls', played specifically to the idea that college-educated women would 'adapt' to their male environment, just as Darwin taught. One of the reasons that Emily Davies encouraged her students at Girton to take the mathematics tripos was because this was the most prized degree for men and a subject generally held to be beyond the capabilities of women. When her students beat the men at mathematics it added ammunition to her argument for intellectual equality between the sexes.

This chapter aims to examine the Cambridge mathematics tripos with reference to women's participation. This will involve presenting the social context within which women approached mathematical study and exploring issues surrounding the tripos training regime that, prescribed for male students, could be ambivalent for women: rigorous mental and bodily training, competition, and close study relationships with men. As well as mathematics having an impact on women and notions of femininity, women's participation in mathematics had significant influence on the tripos itself and affected its relationship with the natural sciences tripos. These latter concerns will be addressed in more detail in Chapter 6, here the emphasis will be on the experiences of female mathematics students, the social context of those experiences, and the negotiations that were necessary to accommodate women (to a certain extent) within the mathematical life of the University. It will be demonstrated that women were not coached in the same way as men, that gendered notions of success were applied to students according to their sex, and that women's performance in the tripos adversely affected the prestige of the examination and contributed to reforms introduced in the later years of the nineteenth century. This chapter will also serve as an introduction to the two women whose lives provide a guiding window through which to view gender, mathematics and science around 1900. Both Hertha Ayrton (1854–1923) and Grace Chisholm Young (1868–1944) studied for the Cambridge mathematics tripos at Girton College in succeeding decades. Hertha became a well-known physicist and electrical engineer; Grace made her life work research in pure mathematics.

Hertha Ayrton

Hertha Ayrton, then Sarah Phoebe Marks, entered Girton College in October 1876, supported by the financial generosity of women's rights campaigner Barbara Bodichon and her circle of feminist friends.[8] She had lived a respectable but impoverished life with her Jewish immigrant family near Portsmouth before moving to London, at around nine years of age, to be educated alongside her cousins in the small school run by her better-connected aunt, Marion Hartog. According to her friend and biographer, Evelyn Sharp, Hertha first became interested in mathematics through the example of her elder cousin Numa Hartog. Hartog, who had attended Trinity College, was the first Jew to attain senior wrangler status in the mathematical tripos (1869) and was admitted to his degree without having to take the usual religious oath.[9] Since receiving political rights in 1858, many of England's Jews had attained positions of status and with her move to London Hertha became part of a comfortable Jewish community which included the family of Sir Francis Goldsmid, the first Jew to become a barrister and an MP. Goldsmid was a financial supporter of Girton College (when Grace Chisholm entered Girton in 1889 she did so as Francis Goldsmid Scholar) and it was probably through this connection that Hertha became aware of the opportunities available there. Despite acceptance of those Jews who had become assimilated into middle-class society, anti-Semitism was on the rise in the late 1870s onwards, ostensibly in response to the arrival of poorer refugees from Eastern Europe. The coming of these immigrant groups coincided with the development of evolutionary approaches to science which codified and reified racial differences, creating a hierarchy of racial groups.[10] In 1905, the Government responded to escalating fears of an 'influx' with an 'Aliens Act' that limited Jewish immigration by excluding those without financial support or suspected of having a 'bad character'. Subliminal fear of the Jewish 'other' has even been implicated in the genesis of the vampire genre, which experienced a resurrection around turn of century, and it has been shown that representations of 'Dracula' and the anti-Semite's 'Jew' are strikingly similar.[11] Whatever the reality of this, it is clear that the new 'Jewish aristocracy' of which Hertha was on the periphery could, by its very conspicuousness, become a potential focus for hostility.[12]

Hertha was the first Jewish woman to attend Cambridge and there is some evidence that her Jewish origins were an issue for her, both at Girton and during the start of her career. Hertha had been rejected initially by Emily Davies, probably in part because of the latter's Anglican leanings. At the time when Hertha entered Girton the celibacy restriction on fellows was still in place (repealed in 1882) and colleges required (male) students to attend chapel. Cambridge was still a firm part of the Anglican establishment and Davies wished Girton to be run on exactly the same lines as the men's colleges; when the new college buildings were constructed in 1902 she insisted on the inclusion of a chapel, despite objections from some of her secularising supporters.[13] Hertha later wrote that at college religion was the great divide with the evangelicals ranged against the anti-clericals and a fellow student recalls her independence and disregard as to whether people liked her or not.[14] Hertha's 'bush' of thick, black, curly hair was an attribute coded Semitic and was used as a defining feature by others. Even her patron, Barbara Bodichon, advised that when she went out to teach she should put her hair in a net as 'it would be worth £50 a year to you'.[15] As well as representative of Jewishness, long, thick, dark hair was also a cultural symbol of wild, instinctive femininity and these kinds of representations jarred with Hertha's later efforts to present herself as a rational, empirical scientist no different from her male colleagues. Hertha's dark looks would often feature in reports of her public lectures and may have enhanced her attraction for the press, at least in the early years of her career, as it added to her exoticness as a female scientist. When she read a paper before an audience of men at the Institution of Electrical Engineers Hertha was described as a little dark-haired, dark-eyed lady who 'created a sensation'.[16] In similar vein, the memoir of a fellow Girton student describes Hertha as 'A poetic and romantic figure, with piercing dark eyes, and wonderful hair ... whose deep voice had extraordinary cadences ... she might well have been the heroine of a story'.[17]

The second Jewish woman to attend Cambridge University was writer Amy Levy (1861–1889) who attended Newnham College and committed suicide at a young age. In her novels and short stories, Levy explored ideas of Jewish self-hatred. In 1886 she published an essay on *Middle Class Jewish Women of Today* asserting that, if they wished to pursue interests beyond the home, Jewish women had to break ties with their religion and its notions of family, race and the importance of marriage. Hertha is named as a high-achieving Jewish woman who traded broken ties for career success by renouncing her religion and marrying a gentile, the electrical engineer William Edward Ayrton.[18] Levy knew Hertha personally; they were friends and fellow members of the University Women's Club, so Levy's opinion on Hertha's choice to renounce Judaism was an informed one. That Hertha seems to have rejected participation in any of the Jewish philanthropic movements that proliferated at the end of the nineteenth century, despite her involvement in many other causes and committees, adds weight to Levy's view. It is also strengthened by a fictionalised account of Hertha's life by her stepdaughter, Edith Ayrton Zangwill, which makes no mention of Hertha's Jewishness and implies a Christian faith.[19] It seems clear that Hertha did not embrace a Jewish identity and that breaking away from Jewish custom helped her in the feminist ways suggested by Levy; it also assisted her assimilation into middle-class, scientific society.

The completion of Hertha's change of name from Sarah Phoebe Marks seems to have taken place upon her marriage to William Ayrton – up until then she was still calling herself Sarah Marks on college records and on patent applications. The adoption of a new first name is usually attributed (perhaps with some romanticisation) to the influence of her friend Ottilie Blind. Blind is said to have given Hertha this new name after a Swinburne poem and because she resembled the Teutonic goddess Erda.[20] Why Hertha chose to adopt this name as a public identity (instead of as an affectionate name used by friends) is significant. The change underlined her desire to break away from, or at least render nominally invisible, her Jewish past. It may also be interpreted as representing a rejection of religious modes of explanation in general in favour of a new code of scientific rationalism (Sarah was the biblical wife of Abraham and mother of Isaac). Later in life, Hertha termed herself an agnostic which she claimed was the 'scientific' approach to religion as one 'cannot say that these things are not true, only that they have not been scientifically proved'.[21] In one of the few extant photographs of Hertha, a 1906 portrait of her in her home laboratory (a formal portrait that was carefully posed to commemorate her winning a Royal Society medal) a painting with a Christian theme is clearly visible in the top right hand corner.[22] (Figure 5.2)

Grace Chisholm Young

By the time that Grace Emily Chisholm arrived at Girton in 1889 the College was fairly well established with an increasing number of women seeking admission. A scheme of extension had commenced in 1886 and this had resulted in twenty-seven new student rooms, bringing the total to 104. In Hertha's time facilities had been more modest, for example a library was not established at Girton until 1884. Relations with the University had developed too and now women were allowed access to examinations on a more formal basis and, in mathematics, were ranked alongside the men on the order of merit. Hertha had sat the mathematics tripos in 1880 when women were only allowed to do so unofficially and had to rely on sympathetic male dons to send them the papers. Had Hertha taken her examination a year later she would have been able to apply for a degree in 1921 when Cambridge began awarding women titular degrees only without membership of the University or voting rights; in 1948 Cambridge finally allowed women degrees on the same basis as men.

Grace was twenty-one years old when she first entered Girton, the average age of entry of women to Cambridge prior to 1900. She was leaving an affluent, middle class, privileged family; her childhood had been spent in Hazlemere, Surrey, where she had been acquainted with the Tennysons and William Morris. Grace had been educated at home by her mother, née Anna Louisa Bell, who had taught all her children Latin, mathematics and music from an early age. Grace's elder brother Hugh had studied classics at Oxford, while her elder sister, Helen, was fragile having been disabled by polio as a child. In her romanticised autobiographical jottings, Grace presents herself as precocious from an early age, outshining her siblings and becoming the favourite of her father, Henry Chisholm, a gifted mathematician who had until retirement applied his talents to Weights and Measures in his Civil Service post as Warden of the Standards. Grace idolised her father as an intelligent, affectionate man who encouraged her mathematical interests, designed three-dimensional models with her and introduced her to geometry. Grace's mother and aunt were sympathetic to women's issues and the family were known to Emily Davies who arranged for Grace to try again for a scholarship after she had under-performed in the Cambridge Senior Examinations due to illness.[23]

Different women, similar choices

Why did these two women from such different backgrounds aspire to Girton College? Martha Vicinus has demonstrated that from the beginning the women's colleges included a relatively wide range of young women from various levels of society. An obvious distinction was between those from wealthy families who attended for the sake of learning, and those who came in order to qualify for a better teaching post.[24] Grace wrote in her memoirs that she went to Girton because she had 'visions of intellectual cloisters like Plato's Athens' and yearned to meet 'the men of intellect of Cambridge' (when she finally arrived at Girton she still preferred the 'intellectual men' to the 'childish girls around her').[25] This romanticism and adulation of intellect as inherently male came to inform the culture of pure mathematics and was, for Grace, a factor in her later devotion to the discipline. Love of learning may have been just as important to Hertha, but her first need was to support herself. She had been working as a governess in London prior to her application to Girton College and, having found being a resident governess onerous, began to attract her own students prior to commencing her studies. Had they been students simultaneously, it is unlikely that Hertha and Grace would have been close friends. Despite attempts by Emily Davies to equalise conditions amongst students,[26] both Hertha and Grace report class divisions. For Grace, the influx of girls from the new high schools 'lowered the intellectual tone';[27] for Hertha, that some students looked down upon one of their fellows because of her shop-keeping background was a reason to proclaim her own origin from the rooftops.[28]

Grace achieved the equivalent of a first-class pass in part one of the mathematics tripos in 1892; she returned to score highly in the more specialised part two the following year, an advanced examination usually taken in a student's fourth and final year at college. In response to a challenge from her brother Hugh who wanted to prove the superiority of an Oxford education, after their success in part one of the tripos, Grace and fellow student Isabel Maddison[29] sat for the final Honours School of Mathematics at Oxford where, according to family legend at least, Grace obtained the highest mark for all students at Oxford that year. Grace's participation in the Oxford examination was purely an informal arrangement and her achievement did not enter the record book.[30] However the fact that Grace and her fellow student were allowed to sit the examination is indicative of women's marginal (even inconsequential) status at Oxford and Cambridge at this time.

Hertha gained a disappointing third in the mathematics tripos in 1880 and did not go on to part two as this specialist extension examination was not available at that time. Like many of the early students, she was ill prepared for university as she had not benefited from the preparatory education, similar to that available in boys' public schools, which was beginning to be offered by the end of the following decade in the new high schools for girls. The growth of these schools was in part fuelled by the growing availability of teachers from the new women's colleges. Hertha had spent time away from Girton due to illness and, while there, preferred to study in the morning and spend the afternoons engaged in more practical work. While still a student she had devised a sphygmometer (device for measuring the pulse) which was to be the first of several inventions for which she took out patents during her life.[31] Given these leanings towards practical science and technology, why did Hertha not choose to study natural sciences in preference to mathematics? The answer lies in the culture of learning at Girton and the perceived importance of mathematics to the campaign for higher education for women.

Special significance of mathematics for women

The combination of mental and physical excellence that success in the mathematics tripos was believed to demonstrate, plus its acknowledged connections to elite masculinity, made the discipline a target for campaigners such as Emily Davies in their strategy to demonstrate women's intellectual equality with men. The culture at Girton, informed by this desire, privileged mathematics over other disciplines; if any student showed a talent for it she was firmly discouraged from dissipating her energies on any other subject. Sara Burstall was dissuaded from taking history because of the mathematical aptitude she exhibited in the early May examinations and, '… it being then still specially desirable that women should prove they could be Wranglers', she was given additional scholarship funding to enable her to spend the vacation preparing for the tripos. Unfortunately she just missed out on a first-class pass, although 'The College authorities were very kind about the disappointment, and the Cause was strong enough to do without a woman Wrangler that year'.[32]

Although this privileging of mathematics had lessened to an extent by the time that Grace arrived at Girton in the early 1890s, the mathematics tripos was still regarded as somewhat 'special'. Such was its reputation that any woman perceived to be on course to achieve a first-class pass was said to have a certain 'glamour' about her. Grace's autobiographical notes and letters also convey the (sometimes none too sisterly) competition between Girton and Newnham to produce wranglers. Newnham scored a high profile win with the outstanding performance of Philippa Fawcett, a success celebrated with genuine and high-spirited jubilation by staff and students at both colleges alike. Fawcett, daughter of Newnham pioneers Millicent and Henry Fawcett, was placed 'above the senior wrangler' in the 1890 tripos; her success a testimony to the significance of mathematics to the early campaigners for women's higher education. Philippa's mathematical talent had been recognised when she was a child since when she had been tutored with the Cambridge mathematical tripos in mind. From the age of fifteen, her well-connected family arranged special coaching provided by a don from Trinity Hall, Cambridge; Philippa then attended courses at Bedford College and University College London. She was therefore, unusually for a woman, as well prepared as the men and this made her an acceptable proposition for her Cambridge coach, the well-respected Ernest Hobson.[33] (Indeed, a fellow classmate at University College was Geoffrey Bennett, the senior wrangler of 1890, a man whose fate was to be forever after linked to Fawcett's name.)[34]

The news that a woman had beaten the 'senior wrangler' made headlines in the local and national press, with the *Telegraph* suggesting that there was now no longer any field of learning in which the lady student does not excel as 'Miss Fawcett has added the last, and possibly most coveted laurel wreath to grace the lofty brows of womanhood'. The report continues (with some ambiguity and a compliment cloaked in irony) to make another implied reference to mathematics:

> We are more than gratified by this result because it removes from our minds one of those lingering doubts which have sometimes interfered with the full and frank admission of feminine superiority.

Across the Atlantic too, the *New York Times* ran an article several columns long.[35] The publicity habitually given in national and provincial media to the results of the Cambridge mathematics tripos was another factor in making women wranglers of particular importance to the 'Cause'. With profiles and pictures of the top wranglers and their coaches commonplace, when a woman was highly placed alongside (or above) the men in this difficult discipline, a message of feminine intellectual excellence was sent out loud and clear.

Mathematics and femininity

But there were other reasons why mathematics could be viewed as an appropriate discipline for women in addition to its power for demonstrating that intelligence was not solely the preserve of the male sex. Women's campaign to make their presence felt coincided with changes at Cambridge that enabled students to transfer directly to the advanced part of the natural sciences after part I of the mathematical tripos. The former also increased its mathematical component and began to rival mathematics as a key route into laboratory-based science – a route increasingly popular with men. As will be addressed in Chapter 6, by end-of-century so many men were taking advantage of this that the decreasing number of male candidates taking the specialist part II of the mathematics tripos became a serious concern to Cambridge's pure mathematicians. This trend helped to inform a new impression of mathematics as more at ease with femininity, both in contrast to its history and to the natural sciences. This 'feminisation' of mathematics was further emphasised by a 'masculinisation' of the more worldly, practical sciences and a contrasting understanding of mathematics as more 'passive' and suitable for women. By its very nature, mathematics was clean, sedentary, safe (unlike the laboratory) and (mostly) removed from the grim and sometimes immoral realities of the real world. In the 1890s a questionnaire was circulated to Cambridge lecturers canvassing their opinion on opening lectures to women. Respondents in classics and the natural sciences complained that the subject matter of lectures had to be modified if given in the presence of ladies.[36] Earlier, the Vice Chancellor of the University had argued that the study of Greek authors was 'bad enough for men, let alone women'.[37] Here the very abstraction of mathematics could be perceived as of benefit to women, rather than an obstacle. This perception was reinforced by the use of symbolic language which ensured the discipline's remoteness from unpleasant aspects of the world, preserving innocence and purity. Mathematical notation may also relieve women of the need to assume uncomfortable and 'unwomanly' authority in their use of language; they could take refuge in impersonal, supposedly objective, symbols instead. No wonder Cambridge women targeted mathematics in their campaign for intellectual equality with men. Paradoxically, it was viewed as the elite 'masculine degree', yet it could also preserve a woman's femininity, which other more 'knowing' disciplines may threaten.

Despite aspects of mathematics that could make the subject a comfortable choice for women, there was still much debate as to whether the discipline was too hard for the female sex or detrimental to femininity and health. Even women such as Sara Burstall, a mathematician herself and firmly committed to girls' and women's education, could be ambivalent to mathematics' place within female programmes of study. Echoing Herbert Spencer's principle of the 'conservation of energy', she argued in 1912 that teaching girls mathematics required too much teaching for too little return, consequently 'We ought to recognise that the average girl has a natural disability for Mathematics. One cause may be that she has less vital energy to spare ....'[38] Spencer believed that female evolution, meaning *intellectual* evolution, had stopped at a stage before man's in order to preserve vital organs for childbirth. If a woman undertook rigorous 'brain work' such as mathematics, energy could be diverted from her reproductive system, threatening fertility and general well-being. Informed by these Darwinian ideas, Burstall advised that mathematical study should be kept at a minimum for girls, not least because of the 'hardening influence' it may have on their femininity.[39] Mathematics, she seemed to be implying, with its logical processes and cold unemotional reasoning, may also tempt girls to subvert their 'emotional' nature and neglect feminine ideals of service. These opinions were shared by the mathematics mistress of Roedean who argued in the *Association of Mathematics Teachers' Journal* that mathematics was of little practical use and 'too difficult for the average girl'.[40]

Some of the most strident voices against women's higher education were those of the medical profession making a case from science for its detrimental effects on the 'less robust' sex. Particular concern was caused by the medical theory of 'menstrual disability', a belief that spawned a condition coined 'anorexia scholastica' which was believed to be a debilitating thinness and weakness resulting from too much mental stimulus, especially during menstruation.[41] Pioneers of higher education took these warnings seriously. When Henry Maudsley published his oft-quoted 'Sex in Mind and Education' in 1874, arguing that women would suffer immense harm to their health by following study regimes similar to men's, Emily Davies and her group were worried that it could hurt their plans as 'there is much truth in it'.[42] They were moved to action when William Withers Moore made similar arguments in an 1886 presidential address to the British Medical Association prompting calls for protective legislation for women of the educated classes analogous to that introduced for women working in factories and mines. In response, women's colleges at Oxford and Cambridge carried out joint research on the health, marriage and childbirth patterns of former students. Their findings contradicted medical opinion in concluding that college-educated women were healthier and less likely to have childless marriages than their less-educated sisters and cousins.[43] Another way in which educational pioneers sought to counter health fears, and to reassure female students and their families, was to include medical facilities within women's colleges, as was done as part of Girton's expansion in 1876. Equivalent facilities were not considered essential at men's colleges.[44] The physical and psychological robustness of women was an issue addressed habitually in references produced by colleges for their female students. For example, when Grace applied for a Fellowship at Cornell University, Girton's Mistress Elizabeth Welsh felt the need to stress her 'great vigour and energy both physically and mentally'.[45]

Negative views about higher education for women were by no means universal however, and as significant numbers of female students succeeded at college with no ill effects such arguments began to lose their credibility and power to cause alarm. In 1884 Edwin Abbot, a headmaster and strong supporter of female education, published a mathematical social satire in which he parodied the resistance to women's education and the tenets of social Darwinism. Abbot's aim was to challenge what he saw as injustice (or sexism, to use a modern term) in contemporary society by taking such views to their logical extreme and so illustrating their absurdity. Based on the notion of the fourth dimension derived from non-Euclidean geometry, in *Flatland* women were represented as straight lines because they were incapable of education or rational thought; men were represented as geometrical shapes with sides and angles – the more sides they had, the higher their social and evolutionary standing. Women had no angles because they had no brain power and were, therefore, inferior to even the lowest of the men – an isosceles triangle. *Flatland* proved very popular and many editions were produced into the next century.[46]

Women's access to coaching

To compete successfully in the mathematics tripos candidates needed to be adept at rapid problem solving and drilled ruthlessly in memorisation and examination technique. These were all skills for which a private coach was essential. Coaches had been the most important teachers at Cambridge at mid-century; although their importance had declined in other disciplines, they were still indispensable in mathematics due to the continuance of the order of merit. It was the College's prerogative to appoint coaches, although the women could express a preference among the small pool of tutors who would consent to take on the role. The College acted as an intermediary and accounting point and, in this way, saved women from the social awkwardness of having to arrange a financial transaction. Receiving payment for intellectual work could still be cause for embarrassment and even male students and coaches may suffer anxiety during the collection of fees. For some coaches, the student was required to hide payment somewhere in the room so that both student and coach could communicate as if no financial exchange had taken place.[47]

The reputation of a coach (and the fees that he could command) was measured by the performance of his students, drawn from varying colleges, and their ranking on the pass lists. Coaches wanted students with high potential; students wanted coaches with a record of drilling high-placed candidates. There is evidence that women were not attractive as students to coaches who sought to maximise their reputations. A coach's standing was dependent upon the performance of his pupils; that the women were often starting from a lesser state of preparedness than their male counterparts compounded coaches' fears that they may not do well.[48] Sara Burstall remembered that

> We mathematical people were perhaps the worst off. All our work was private coaching with such Cambridge men as could be induced to come out three times a week, nearly two miles of dull walk along the Huntingdon road, to take on the teaching of young women, often ill-prepared and unlikely to reflect honour on their teachers.[49]

One leading coach of the time, Edward Routh, declined to train Charlotte Angas Scott who went on to be placed equal to the eighth wrangler in 1880 (unofficially as this was prior to women's inclusion on the order of merit).[50] According to Hertha, who was her fellow student at Girton, Angas Scott was third on the pass lists after the first three days of examinations but dropped position later due to 'not having read enough, the result of having read so very little before she came up ...'.[51] Even Philippa Fawcett, who had been trained as well as the best men and marked out as a high achiever, had been refused permission by another top coach, Robert Webb, to attend his classes. Such was Webb's antagonism towards women that he maintained that if Fawcett beat his candidate, Geoffrey T. Bennett of St. John's College, both he and his pupil would emigrate to the new University of Chicago.[52]

Webb's refusal to coach Fawcett had more to do with propriety and concerns about having a woman in the coaching room than any fears about her ability. One of Webb's male pupils remembered that he had had 'a rough tongue' and that he had refused Fawcett because 'he considered that the presence of a lady in his classes would prevent that freedom of language necessary for teaching mathematics'.[53] The environment of the coaching room was rough, individualistic, competitive and frank – not a place where the polite conventions and middle-class etiquette required for social intercourse between the sexes could be maintained. The coaching room was simply not a place for a 'lady' and, moreover, her presence could cause unease, discomfort and concern about correct behaviour amongst the men. As late as 1911 a book entitled *The Intellectual Life* had testified to the difficulty of authentic communication between the sexes due to the requirements of etiquette, requirements which were 'hereditary and instinctive' for men. These rules made it 'quite impossible for men to speak to ladies in the manner which would be intellectually most profitable to them' as 'we may not contradict because it is rude'. The author concludes that 'Men will never talk to women with that rough frankness which they use between themselves. Conversation between the sexes will always be partially insincere'.[54] Concerns such as these all too often resulted in women's exclusion from the coaching room of the top coaches who typically ran large and regimented training regimes for their male pupils.

Another outcome of coaches' reluctance to teach women was that it became very difficult to maintain both the quality and continuity of coaching for Girton and Newnham's mathematicians. Sara Burstall was trained by various coaches for 'short periods' and attributed her failure to achieve a first-class pass, in part, to the shortcomings of the 'inexperienced coach' assigned to her prior to the examination.[55] The men who lectured and coached the women, unsurprisingly, tended to be supporters of women's higher education and, often, critics of the Cambridge system too. Grace was tutored mainly by Arthur Berry of Kings (1862–1929) who served on the Executive Council of Girton College and who, as Secretary of the University Extension Syndicate, moved to allow women lecturers in 1893. Arthur Cayley was active in the foundation of the 1869 lecture series for women which eventually led to Newnham College; he taught many of the early students including Charlotte Angas Scott.[56] Angas Scott attended Cayley's lectures in the 1880s and he was also instrumental in opening Grace's eyes to mathematics beyond the tripos by welcoming her, and a fellow student, into his home and taking them to a special lecture he was giving in Cambridge. This was unusual, especially for women. There was a huge gulf between undergraduates and professors and the former rarely attended College lectures as they were largely irrelevant for part I tripos examinations, becoming important only for the specialist part II. Students were required to pay extra for attending these events, but the less well off women could sometimes dispense with this expense by attending lectures as a chaperone – something Angas Scott, daughter of a Congregational minister, took advantage of.[57]

Hertha, who sat the mathematics tripos in 1880, received the fragmented training that was typical for early students. She was taught at one stage by Richard Glazebrook who had also tutored Burstall for a while. Despite being fifth wrangler in 1876, Glazebrook was critical of the tripos system, particularly its separation of mathematics from experimental work and the continuance of the order of merit. At the time he was coaching the women he was working as a demonstrator at Cambridge's Cavendish Laboratory and, given his preference for experimentation, Glazebrook was not considered by ambitious candidates as a coach likely to push them towards a top place on the list. Like Glazebrook, many of the coaches and lecturers who visited the women's colleges came with a reform agenda which made them hesitant simply to replicate the education that was given to the men. These 'youthful and enthusiastic young gentlemen ... were more concerned to advance general culture than to coach for exams' and this led to protest from Emily Davies at Girton and some of her early students.[58] One of those students was Grace Chisholm who wrote a letter of complaint about one young mathematics lecturer, stressing his youth, inexperience and 'lack of seriousness'.[59]

One answer to these problems was to recruit women into the coaching role and there are suggestions that parallel systems were seen as a solution to questions of propriety as well as of access. After her high placement on the pass list, Charlotte Angas Scott was told by one of her examiners that 'if she would stay in Cambridge she should have his sister to coach at once'.[60] Scott did remain as a lecturer at Girton for a short while, as did Fawcett at Newnham for nearly nine years. However, despite reforms to university teaching that lessened the power of coaches in the latter decade of the nineteenth century, there remained a mythology surrounding the most celebrated coaches, usually based on their eccentricity and the robustness of their teaching techniques, that the female tutors found hard to emulate. The position of the female don was not of an equal status to her male counterpart: she tended to be poorly paid, isolated and with little say in the curriculum or governance of the University. For Grace, even in the 1890s, 'nearly all the head lecturers are men and, as for the female dons, they are chiefly there to quiet the anxiety of parents for their daughters and act as chaperones. Nobody with any pretensions coaches with them if they can help it'.[61] Coaches won work according to their reputation, based in part on the number of wranglers they coached; it was difficult therefore for female tutors, teaching comparatively few less-prepared students and with pastoral duties to perform as well, to compete in this arena. Fairly or not, the reputation of some female mathematics tutors was not high – Grace's notes also record rumours circulating that the coaching offered at Newnham by 'Miss Fawcett' was not good.[62]

There is no doubt that the long-held associations of the mathematics tripos as a vehicle for testing manliness as well as mathematical skill posed a barrier to women being accepted as effective coaches. However this association was in the process of being eroded in the last two decades of the nineteenth century as increasing numbers of women were successful in mathematics, giving that examination a more feminine colouring while the mantle of masculinity passed to the newer natural sciences tripos. Despite this (and Davies' desire to see women coached to compete successfully with the men) there is evidence of tension in the coaching room between female students and their male tutors and signs that the teaching techniques used for women were different from those used for men.

Tensions in the coaching room

There are many accounts left by male wranglers describing their experiences in the coaching room, both in memoirs and obituaries, which are alike in recalling the immense hard work required and the relentless need to compete with one's peers. These accounts collectively comprise a shared mythology, with communal terms of reference, which was used as a model and added to by succeeding generations. A.R. Forsyth sat the tripos in 1881, the same year as Sara Burstall, but his memories of being coached are very different. Forsyth was trained by the legendary coach Edward Routh[63] – an experience which he described as 'a marvel even of physical endurance, let alone intellectual effort'. Routh's 'system' was to offer one-hour classes, three times a week, on alternate days during term time and the long vacation. Classes were attended by a crowd of some twenty ambitious young men who were required to complete exercises and solve problems between sessions, their answers graded and displayed publicly. Teaching was devoted entirely to how to frame an examination question as speedily as possible and this involved 'scribbling hard ... not a moment spent in diversion or extraneous illustration ... there [was] little leisure for thinking, because we were all being taught'.[64] In similar vein, Robert Webb laid claim to practically all the time and energy of his pupils and could be harsh with students who failed to meet his exacting demands.[65]

Women's memories of being coached seldom convey this sense of relentless pressure, or any indication of sharp words or recriminations. Unlike Routh's industrialisation of the coaching process with some twenty men in the coaching room, women were generally taught in pairs in a more gentle style. For the students at Girton days were highly structured and the avoidance of mental strain and undue competition was paramount. Hertha was restricted to five hours study a day[66] and Grace expected to attend just one lecture each day, in the morning or afternoon, completing no more than six hours work with nothing after 6pm.[67] She also records receiving one hour's coaching a day, three days a week. Control, not work to the point of exhaustion, was similarly a hallmark of Fawcett's methodical study regime. This was composed of 'six hours work, *very* rarely exceeded, plenty of regular exercise and always to bed at 11.0 (sic)'.[68] Coaches were wary of pushing the women as hard as the men: while Forsyth records being coached at a 'wonderful pace … not a moment was wasted … (just) grim doggedness and unresting drill',[69] Grace recalls a more leisurely, although hardworking, pattern that included a break halfway through when the tea tray came in and she poured.[70] Grace was given problems to solve between lectures, but she did not experience the pressure of repeated, public competition against peers in mock examination papers (as was Forsyth's lot).

Relationships between coaches and the women they taught could be uneasy; Girton and Newnham's students tended to be taught by younger, 'greener' and less experienced fellows who may have only just sat the mathematics tripos themselves. Many of these young men were not much older than their pupils (for example Hertha was exactly the same age as her coach Richard Glazebrook) and, for women and men, being in such close proximity to a member of the opposite sex, especially in an academic context, could be unsettling. Although Grace's autobiographical account of her Girton years is romanticised and written with hindsight, her identification of a strong thread of novelty and tension within the female student/male coach relationship is plausible:

> It was quite a new experience … to come into contact with these male lecturers. She had danced and played tennis with young men of her own age, but here she was seated at a long table with a young man, crammed full of that mathematical knowledge for which she thirsted, and who poured it out for her, at the end of a quill pen, without any touch of familiarity, for the space of an hour, three times a week.[71]

In addition to avoiding 'familiarity', the 'young men' had other strategies for coping with a difficult situation: one walked straight into the lecture room at Girton and, without any greeting or acknowledgement, proceeded straightaway with his lecture. Interaction such as asking questions was frowned upon, it could threaten the formality of the proceedings and it held up the class. If students got lost or did not understand, they asked a fellow student.[72] This strict sexual protocol could result in episodes bordering on the surreal. Beatrice Cave-Brown-Cave of Newnham College recalled a coaching session with William Young (who briefly coached Grace and later became her husband) during which he kept tilting his chair until it slipped and he went under the table. With great difficulty Beatrice and her fellow student refrained from laughing as Young righted himself, ignored what had happened, and merely told them to take out a fresh sheet of paper.[73]

Reconciling competition with femininity

Competition was a key element of the Cambridge mathematics tripos; it was made manifest in the annual – and very public – celebrations surrounding announcement of the order of merit and it informed the close relationship that was perceived between competitive sport and the examination. For women competition, especially with men, threatened contemporary feminine ideals of modesty, selflessness, domesticity and service. A concern to connect with the domestic, instead of to the assertive and individualistic ideals of male academia, has been identified even in the architecture of Girton. Instead of being built to mirror the men's colleges with their grand, confident, institutional designs, Girton's purpose-built premises relied instead on a domestic architectural model which featured inglenooks, bay windows and roof dormers.[74]

While anticipation of a high place in the mathematics tripos could give a woman the 'glamour' of 'probably a wrangler', there was also ambivalence in the students' attitudes to success and even the most high-achieving female students were praised for 'never displaying their cleverness in the wrong way'.[75] Delivering a paper to the Girton Mathematical Club, Charlotte Angas Scott warned her audience against being tempted to 'win a name for yourselves' and, instead, encouraged them to develop a genuine love for mathematics.[76] Reflecting similar concerns, a contemporary chose humility as Philippa Fawcett's most outstanding characteristic and praised her for being 'modest and retiring, almost to a fault ... so as to appear like a very ordinary person'.[77] But reticence to compete with men was not so evident when it came to competing with each other or against another women's college. Grace's autobiographical notes are full of remarks about fellow students' abilities, while Hertha felt it particularly unjust that Newnham students, unlike Girtonians, were not obliged to adhere to the same examination conditions as the men. To follow the men's programme required passing the Previous Examination (or Little Go) in the first year, which tested Latin, mathematics and Greek, and sitting the tripos examination within three years and one term. As Hertha complained to Barbara Bodichon about the 1879 tripos: 'Newnham has two students in, one has been up four years and hasn't taken her Little Go, so of course she will take the shine out of ours. I think it's horribly unfair ...'.[78]

Women's concern to present themselves as humble and unexceptional can be interpreted as a strategy to counter hostility against them, hostility which increased in relation to women's success. Rita Tullberg suggests that it was only in the 1890s, when women were revealed to be as able as the men, that resentment against them competing with their male peers at Cambridge increased.[79] This bad feeling culminated in the infamous and overwhelming vote by University Members against giving women degrees in 1897. It is also reflected in the differing accounts of Philippa Fawcett's 1890 success. Unlike the celebration which is the hallmark of reports in the women's college magazines, memoirs written by male mathematicians tend to represent her achievement as odd and treat it with amusement and indulgence. Her coach, Ernest Hobson, is remembered in an obituary as having 'enjoyed one theatrical triumph' by having the female senior wrangler as one of his pupils.[80] A later account by Grace's son, possibly coloured by loyalty to his mother, recalls that this 'daughter of a radical economist and of a most militant feminist ... achieved unique fame in the annals of feminism' and 'ruined the life' of Bennett and his coach.[81] This sceptical indulgence of women's intellectual ambitions would have been recognised by Charlotte Angas Scott; she wrote the following words in 1898 to her principal at Bryn Mawr in the States, but she may well have been reflecting on her Cambridge experiences too:

> I am most disturbed and disappointed at present to find you taking the position that intellectual pursuits must be 'watered down' to make them suitable for women ... I do not expect any of the other members of the faculty to feel this way about it; they, like (nearly) all men that I have known, doubtless take an attitude of toleration, half amused and half kindly, on the whole question; for even where men are willing to help in women's education, it is with an inward reserve of condescension.[82]

The competitive mental exertions required of male students by their coaches was accompanied by a harsh regime of physical drill; both were believed necessary to produce what Warwick has identified as the manly ideal of the elite mathematics student in whom the rational mind and strong body were perfectly combined.[83] Although this ideal was losing some of its influence by the 1880s and 1890s, hard work, competition and regular physical exercise were still felt important to achieving top mathematical honours. However, for women these prescriptions were problematic and clashed with contemporary notions of femininity which accepted that women's bodies and minds were not as robust as men's. When Henry Maudsley warned about the threat posed to women's well-being by hard intellectual toil, Dr. Elizabeth Garrett Anderson in reply did not deny that nervous breakdowns and ill health were genuine problems; instead she argued that they were caused by a *lack* of mental stimulation, not too much. She then reassured that steps were being taken by educational reformers to guard against these problems and develop girls physically, by which she meant that controlled study was complemented by a programme of exercise.[84] However, whereas strenuous physical exercise was deemed desirable for male mathematicians to prepare *for* the strain of the tripos, women took gentle physical exercise to guard *against* strain. At first they took up gymnastics because it was thought that this would build up their delicate frames for study. Athleticism increased towards the end of the century, but women were still obliged to show restraint and conform at all times to 'ladylike' behaviour.[85] New activities such as hockey, golf and tennis were taken up with enthusiasm by students at the women's colleges, and it is significant that these were mostly 'domesticated' games which substituted rules, team work and co-operation for the aggressive individualism of running or rowing. Tennis was played by Hertha and Grace, and they both participated in drills of the Girton Fire Brigade – 'the first really masculine piece of organisation devised by Girtonians'.[86] Girton had a golf course in the 1880s; in the 1890s hockey was played, but only sedately, in full dress, away from male eyes.

At the end of the nineteenth century, sport was still overwhelmingly a symbol of masculinity; it was seen as an arena in which to develop courage and competitive instinct, both essential for success in the tripos. However, for male candidates physical failure in the face of the examination could also be interpreted as a sign of asceticism, abstraction and increased intellectuality. Warwick has written of 'funking fits' (collapses in the examination room) and notes that as early as mid-century being pale and ill could be a sign of intellectual strength.[87] Moreover, the 'genius' of leading scientific and mathematical figures of the late nineteenth and early twentieth centuries was often contextualised by contemporaries in similar ways. Rayleigh recalled that J.J. Thomson could not stand up to the day after day physical strain of the tripos[88] and G.H. Hardy remembers missing out on showing his talent for sport at school as 'no one thought it worth looking for in the school's top scholar, so frail and sickly, so defensively shy'.[89] It is suggestive of the power of gender stereotypes that failure to stand up to the rigours of the tripos or competitive sport could be interpreted as a sign of intellectual strength among male mathematicians, while for women a purported lack of physical stamina was interpreted as symbolic of precisely the opposite. Both Hertha and Grace experienced bouts of ill health and headaches while students; by the end of the nineteenth century women were constantly defined by their reproductive bodies and for university women, singled out for their minds, there was a special tension that conflicted the requirements of rationality and femininity. Grace experienced this tension acutely, believing that she had 'had a certain career in the University world, and have managed to be one of the few women who do so without sacrificing health'.[90]

Gendered notions of success

Just as a failure of nerve or health was interpreted differently according to whether it was manifested by a man or a woman, so success in the tripos became associated with different, gendered explanations. As women became increasingly visible by winning high places on the order of merit, wrangler status became linked to a student's capacity for hard work but lack of originality; conversely, failure of men to achieve a top place could be rationalised away as indicative of mathematical creativity and a marked potential for research.[91] Similarly, originality in a mathematician was underscored by stressing that he 'never gave a thought to the tripos'.[92] The examination itself, with its requirement for wide knowledge, speedy problem solving and the memorisation of formulae and model answers, became indicative of hard work and dull minds – a characterisation particularly aimed at women. This criticism became more endemic after the introduction of the specialist mathematics extension examination (part III, later to become part II) in the early 1880s which, for the first time, allowed students to choose a narrow area to study in depth.

The idea that women were 'faithful followers', 'diligent' and 'paid meticulous attention to details' but were 'not capable of great creative work' had been a well-rehearsed argument since the 1870s.[93] This commonly-held assumption that women worked harder than the men to achieve their results, making 'the air' at the women's colleges 'tense', reflects this characterisation of women as conscientious but not original.[94] A testimonial for Sara Burstall, in support of her application to teach mathematics at Manchester High School for Girls, exhibits a similar subtext in its praise of her 'unwearying power of work ... industry, intelligence and enthusiasm'.[95] Such views were summed-up succinctly by William Henry Young (later husband of Grace Chisholm) when he argued in a 1913 critique of late nineteenth century Cambridge that the fact that a woman (Fawcett) had already succeeded in beating the senior wrangler had destroyed the prestige of this award and contributed to the abolition of the competitive merit system.[96]

That some women did well in the mathematics tripos, then, helped to devalue the examination rather than to raise the reputation of the women themselves; candidates from Girton and Newnham were understood to have done well due to their diligence, but this did not imply women were capable of original work. To use a modern term, women's success indicated that the tripos had been 'dumbed-down'. In her later years, even Girton wrangler Grace came to find the faults of the tripos 'repulsive' and, revealing a hint of resentment, wrote that 'if you do well at exams then you are not original ... Philippa Fawcett was not, who she beat, Geoffrey Bennett, was ... curiosity about unscheduled mathematics is depravity'.[97]

These conflicting views on the merits of the tripos need to be understood within the context of wider debates about the reform of Cambridge mathematics which will be explored in Chapter 6. However, concern about the worth of the tripos examination system had antecedents from early in the century when the 'objective' ranking of mathematics students was facilitated by Charles Babbage's and John Herschel's introduction of symbolic analysis to Cambridge. This was a technique which attempted to industrialise the thinking process by reducing mathematics to skill in manipulating abstract algebraic notation.[98] While this conception of mathematical intelligence had the potential to democratise thinking by making it a skill that could be acquired, its opponents referred back to a more romantic view of intelligence as a special gift that was inherited, not learned. Although critics, such as William Whewell,[99] affirmed that mathematics should be applied to the world, they feared that abstract 'analytics' was trying to distort that world by forcing it to conform to a mistaken and confining structure which did away with any notion of inspiration or the divine. The arguments over the tripos at the end of century paralleled these earlier debates and also marked the beginning of an inversion of the issues that created antagonism between pure mathematics and applied, constructing each in opposition to the other. It will be argued in Chapter 2 that the association of romanticism with abstract mathematics was a late nineteenth century phenomenon which inverted earlier hierarchies. When a new style of continental analysis was introduced to Cambridge around 1900, its supporters derived this new mathematics' legitimacy and moral currency from its irrelevance to the real world and from the fact that, like Art, it was the product of 'great minds'. Like Whewell before them, pure mathematicians warned that mathematics pursued solely for utility's sake, not for its own, devalues the practitioner and leads to social decay.

A wider context: Oxford and London

Cambridge's unique place in the history of mathematics, especially with reference to its mathematical women, is put into sharp relief by an assessment of the situation at the Universities of Oxford and London.

Oxford University had been less receptive than Cambridge to the idea of women's colleges. According to Janet Howarth, supporters of women's education in Oxford did not seek the role of pioneers and by the time that Somerville and Lady Margaret Hall opened in 1879, with twelve and nine students respectively, over 300 women had already passed through Girton and Newnham colleges.[100] Although Somerville and Lady Margaret Hall were joined by St. Hugh's in 1886 and St. Hilda's in 1893, the number of women students at the Oxford colleges remained small compared to those at Cambridge. This slighter presence was reflected in the lack of formal connections between the women's colleges and the University. Whereas Cambridge had opened mathematics and other examinations (although not degrees) to women on an official basis in the early 1880s, the Oxford halls for women remained independent and unrecognised until 1910. Although women were allowed to sit some examinations from 1884, including mathematics, the University was careful to distance itself from any seeming public endorsement. Technically, the Oxford women were still examined by the 'Delegacy of Local Examinations' which had merely been granted the right to make use of undergraduate examination papers.[101]

For ambitious mathematicians of either sex however, Oxford was not the first choice for study. The University had an especially strong reputation for classics and the humanities and its mathematics school suffered in the shadows; neither could it compete with the towering reputation of the Cambridge mathematics tripos. The most popular subject for women at Oxford was Modern History, followed by English and Modern Languages, with natural sciences and mathematics proving least popular. Whereas women at Girton were steered towards mathematics, at Oxford the situation was reversed. 'Not mathematics, you simply can't come to Oxford and do mathematics' was the exclamation made by Miss Wordsworth of Lady Margaret Hall to one of her new students.[102] Even at Somerville, named for the mathematician Mary Somerville, the situation for ambitious mathematicians was hardly better. It was reported in 1881 that the mathematics students were 'in a terrible commotion' because they were being taught by a professor who was 'lazy, lacks sense, and has no interest in the girls' work'.[103] The lack of status of mathematics at Somerville at this time is also suggested by the experience of Margery Fry who was appointed librarian in 1899 and found herself 'entrusted with a range of tasks from marking hand-towels to coaching in mathematics'.[104] Indeed, figures suggest that in the years 1881–1913 just a handful of Oxford women were successful in examinations of the honours school of mathematics (a comparative figure for Cambridge is 480 women between 1882–1914).[105] However between 1891–1909 many women from Royal Holloway College were successful in the mathematical honours examinations of the University of Oxford, as discussed below. At Oxford itself however, there was not a strong mathematical culture around turn-of-century and this translated into limited opportunity and encouragement for mathematical women.

The University of London is usually credited as the first institution to open its degrees to both sexes on an equal basis (except for medicine) in 1878, but as Carol Dyhouse would caution, the 'pattern of women's admission is complex, and much depends on definitions'.[106] London's evolution was very different to that of its ancient counterparts; it was formed in 1836 as an examining body (offering services predominantly to students of University College and King's College London) and because of this imposed no residential requirement on candidates presenting themselves for examination. This not only benefited women wishing to study independently at home, it also became a route for 'graduates' of Cambridge and Oxford to gain the formal degree qualification denied them by their former institutions, including some women mathematicians.[107] After 1898 the University became a federal 'Teaching University' and began offering courses in its constituent colleges. In 1907 University College London (which in 1878 had also opened degree programmes in the Faculties of Sciences, Arts and Laws to women) ceased to have a separate existence and was incorporated into the University of London. Prior to this, in the years 1898–1900, women's colleges in London and Surrey had been incorporated. The latter comprised Bedford College (established 1849), Westfield (established 1882), Royal Holloway (founded 1879 and opened by Queen Victoria in 1886) and King's College which had offered a 'Ladies' Department' since 1885. However this expanded University of London still retained its role as an examining body and distinguished between 'external' candidates and 'internal' ones drawn from its own colleges or schools. Just a handful of women graduated in the early years, however by 1900 there were 169 women awarded degrees, representing over 30% of the total.[108] The first women mathematicians to graduate were Sophie Bryant (1881) and Charlotte Angas Scott (1882), who sat the examination after her tripos success and, three years later, became the second woman to be awarded a DSc from London.[109] By end of century there had been only eight female mathematics graduates, but by 1914 eighty-six women mathematicians can be identified, plus two women MAs.[110]

The University of London's constituent colleges had a longer history in providing courses and research. At UCL the physical and natural sciences had been dominant since the early days and this continued into the twentieth century, with a scattering of women graduating in subjects such as botany and zoology. (Karl Pearson, following in the footsteps of Francis Galton, set up Biometrics and Eugenics Laboratories at UCL in 1903 and 1907 respectively and employed a handful of women mathematicians as researchers and statisticians, as referenced in Chapter 6). Although UCL had some eminent mathematicians as professors in its early days, including Augustus De Morgan who tutored Ada Lovelace as a private pupil, mathematics was not a popular subject for either sex in the first years of the new century.[111] Elizabeth Larby Williams entered Bedford College in 1911 and remembered that mathematics was not a subject studied by many and in order to provide a sufficiently large audience 'there was a combined programme of lectures for students from Bedford, Kings, University and Westfield colleges. Even so the class was barely twenty strong. Yet there was almost as many women as men.'[112]

However a strong culture of women's mathematics prior to 1914 can be identified at Royal Holloway College which produced over forty women graduates of the University of London between 1903 and 1913.[113] In addition, up to 1909, seventeen women were successful in the Final Honours examination of the Oxford School of Mathematics, with many others passing mathematical 'moderations' (intermediate examinations).[114] This flowering may have been due to the influence of Ethel Maude Rowell who had been an outstanding student at Royal Holloway and went on to teach there from 1899 to 1939, and to other enthusiastic women mathematicians who joined her at various times on the staff. It was also encouraged by Royal Holloway's 'Founders Scholarships' (in 1892 for example, four of these were awarded in mathematics while two were awarded for other subjects) and by annual prizes for both pure and mixed/applied mathematics. Rowell (d. 1951) was a veritable institution at Royal Holloway, having been intimately connected with the College for over forty years. She was a philosopher and poet as well as a mathematician and had been introduced as a child to mental exercise and logical thinking by the Rev. Charles Dodgeson (Lewis Carroll). She remembered him objecting to her plans to study at Royal Holloway for the final honours examination in mathematics at Oxford because 'the work was far too exacting and would impose a strain which may even upset my mental balance!'.[115] Fortunately, this did not deter the young Ethel.

Conclusion

An exploration of the culture of mathematics at Cambridge in the decades around 1900 illustrates that gender was a key determinant in the training provided for students of the mathematics tripos. Although an affinity between mathematics and femininity can be detected towards the end of the nineteenth century, it is clear that studying for the mathematics tripos was a very different experience for women than for men. Not only did the type and quality of coaching given to female students differ from that provided for their male counterparts, but the gendered meanings that became attached to women achieving a high place on the pass list resulted in this 'feminine' success being effectively devalued. This, in turn, had a reciprocal effect on the status of the tripos itself. Towards the end of century, the examination lost much of its prestige and reputation as an elite qualification due (in part) to an increasing recognition that women were able to compete successfully alongside the men and, in some cases, surpass even the best of them.

The culture at Girton among students studying for the mathematics tripos was informed by a pattern of tensions and attitudes that helped set the course for Grace's and Hertha's future careers. Grace's repudiation of tripos competition was key to her later espousal of elitist ideas concerning the male intellect; it also informed her decision to put her mathematical skill at the service of her husband and her belief in the absolute superiority of pure over practical mathematics. Hertha's experiences at Girton a decade earlier were more problematic; her love of experimentation and practical science did not make for an easy life at college at a time when the mathematics tripos, which had little experimental content at the time, was privileged above all others. Grace can be seen as representing the romantic traditions now attached to pure mathematics conceived as an abstract enterprise for gifted individuals; Hertha embraced the newer sciences, such as electrical engineering, which promised new opportunities as they began to challenge older hierarchies and made use of a practical, utilitarian and 'manly' mathematics. These hierarchies also intersected with notions of class: in a culture and educational context in which 'hand' and 'brain' had long been opposed, practitioners of engineering and the more practical sciences sought to raise the status of their professions and challenge the assumptions of superiority of the pure mathematicians.

Hertha's choice of the practical, and Grace's championing of the pure, may have been influenced by their very different backgrounds. It could be argued that Hertha, as daughter of a Jewish-émigré watchmaker, would have acquired none of the prejudice against practical or manual labour that was one of the defining features of the educated, English middle class. The differences in the culture, material practices and gendering of pure mathematics and the practical sciences become more evident as we follow, in the succeeding two chapters, Grace to the 'Shrine of Pure Thought' in Göttingen and Hertha to Finsbury Technical College and the Central Institution at South Kensington. Despite their differences in outlook and choice, both women found themselves caged in by prescriptions of femininity which limited their participation within their chosen disciplines and affected their options, reputations and scientific credibility.

2
Women at the 'Shrine of Pure Thought'

In April 1895, Grace Chisholm Young was 'wonderfully happy'. She had just received an honour that she had determined to 'move heaven and earth' to achieve; a prize which had required this young mathematics wrangler from Girton to travel alone to a foreign country, with little knowledge of the language and culture awaiting her, without even an assurance that she would be given permission to pursue her goal when she arrived. But after much hard work Grace had finally achieved her aim. She had been awarded a doctorate in mathematics from the prestigious University of Göttingen in Germany. Grace was now keen to prove herself; her devotion to mathematics had been 'vindicated' and she was ready to 'take my stand among mathematicians, out of the apprenticeship years, able myself to do work'.[1] And she had already begun. As well as receiving a guinea from the Home Reading Union for a paper on sound, Grace's *Vorträge* (seminar paper) had been accepted for publication by the Royal Astronomical Society[2] and her doctoral thesis was being prepared for the printers right now.

Yet the years ahead did not unfold in the way that this ambitious young mathematician, so confident in her intellectual abilities, had planned. During the next few years Grace struggled to achieve the future that she had hoped for and, finally, she decided to realign her priorities – an adjustment that contributed to a major bout of depression around New Year 1900. This woman, who had resolved to earn an independent living, acquiesced in putting her mathematical skills at the service of her husband and absorbing her mathematical personality into his. Two decades after achieving her doctorate, Grace was arguing that a woman, 'whatever her personal ambitions, really longs for a superior male mind' and yearns for the support of 'the complete man'. As a new 'Doktor' Grace had interpreted her success as a blow for the cause of women's education. She had praised Göttingen for the freedom and equality that it offered, unlike Cambridge, to women like herself. Now she warned that any girl who, 'having been granted entry into a society of intellectual boys, lets her personality intrude itself on their feelings, is sinning against the unwritten law of womanhood'.[3] Grace had achieved her doctorate in a subject generally held to be 'too hard' for women, yet just five years after her success she endorsed the views of Professor Max Runge of the Medical Faculty of Göttingen University who argued from evolutionary theory for women's lesser mental capacities.[4] How was such a transformation of viewpoint and behaviour possible?

In the last chapter evidence of the seeds of a rift between pure and 'physical' mathematics was introduced with reference to late-nineteenth-century criticisms of the Cambridge mathematics tripos. Those seeking to reform that examination pointed to its privileging of repetitive problem solving over creative mathematics and one of their solutions was to introduce a specialist extension examination that enabled candidates to specialise in either pure or applied mathematics. In this chapter the focus will move to the University of Göttingen in Germany, to where many critics of the tripos looked for the new mathematical thinking and research-led structure that they sought to use as a model to revitalise Cambridge. Grace's coach Arthur Berry had spent a year's sabbatical at Göttingen, during which time he had been introduced to the advanced analysis and theory of functions being developed there. His admiration for these new, highly abstract mathematical ideas was shared by other Cambridge reformers such as Ernest Hobson, Andrew Forsyth and, later, Godfrey (G.H.) Hardy.

Following on from the antagonism within mathematics perceived at Cambridge, this chapter will examine the processes by which pure mathematics became configured in opposition to physical or practical mathematics, how its language and terminology became feminised, how ideas of male genius became central to its culture, and how all of these had implications for women mathematicians. The discussion will be preceded by an exploration of Grace's experiences in Göttingen (to where she returned to pursue research and make her home after her student years) which served to create severe tensions within her self-identity. Göttingen was a university town built upon the reputations of its 'great men' of mathematics, a place where the spirit of romanticism persisted and where hero worship of male intellect was one of the dynamics behind the development of the School of Pure Mathematics. Grace's natural elitist politics found their reflection within this community in which male intellectual transcendence was glorified and pure reason, in the cloak of abstract mathematics, was viewed as morally superior. The structure and culture of the mathematics department, and the type of mathematics pursued there, found their counterpoint in contemporary ideals of femininity. These ideals, in Germany even more than elsewhere, were rooted firmly in ideas of marriage and motherhood. Within this context, Grace found herself caged in by prescriptions with which she struggled to the end of her life.

Mathematical choices – the options for women

After receiving her doctorate, Grace returned home to consider the options available to her; she had resolved to remain unmarried and earn her living by mathematics. Had she been a male wrangler, a fellowship at Cambridge would have been likely, especially if she did well in the annual Smith's Prize – an important competition for ambitious Cambridge mathematicians. At this time the Smith's Prize was awarded for an outstanding essay and was open to the best placed wranglers of the year – as long as they were men. Philippa Fawcett was not eligible to compete in 1890, but the career of the man she beat typifies what a top male wrangler may expect. After being classed along with Fawcett in the first division of part II the following year, G.T. Bennett was awarded the Smith's prize in 1892 and became a fellow of both St. John's College Cambridge and University College London, moving from St. John's to Emmanuel the following year.[5] Bennett achieved a DSc, was elected a Fellow of the Royal Society in 1914 and spent his life as a respected college lecturer.[6] Such opportunity was rarely available for a woman, at least not in Cambridge. Grace knew she had to look abroad for better chance of a fellowship.

At first America seemed a more promising proposition and Grace discussed the possibility of a move to Bryn Mawr with her friends Isabel Maddison and Charlotte Angas Scott who were both on the staff of this new women's college.[7] Grace had applied unsuccessfully to Cornell University in the spring of 1893 and now considered an application to Chicago where her fellow doctoral student from Göttingen, May Winston, was based.[8] The campaign for higher education for women was gaining firm ground in the States with new women's and co-educational colleges being founded in the 1890s and access to doctoral degrees accelerating. In 1878, J.J. Sylvester at Johns Hopkins had allowed Christine Ladd (Franklin) to attend his mathematics lectures and, recognising her ability, had awarded her via his department a

$500 year-long fellowship. Yet she was denied a doctorate and her attendance was as a 'special student' setting no precedence for others of her sex.[9] By the early 1890s Bryn Mawr and several co-educational institutions offered doctoral programmes in mathematics open to either sex and by 1900 ten American women had been awarded PhDs in the subject; the first decade of the twentieth century saw another eighteen women join them. Chicago, Cornell and Bryn Mawr produced the highest number of women PhDs before 1914, although the first was awarded by Columbia to Winifred Edgerton in 1886.[10]

The number of women in America studying advanced mathematics was reflected in its relative popularity as an undergraduate option. When the University of Chicago undertook a survey of undergraduate preferences in 1902, it was discovered that 16% of women took mathematics beyond the required level despite the current belief 'that women have no aptitude for mathematics'. At Boston, 9% of women and 11.4% of men chose mathematics as their core study.[11] Despite this, American universities at that time could not offer women, or indeed men, opportunities to study at the most creative level with mathematicians whose work was setting the agenda for research.

Among Americans and Europeans of both sexes, from around 1876 the University of Paris (Sorbonne) became a favoured destination for students of the arts and the sciences and numbers grew steadily throughout the Third Republic. In the Faculty of Science the student population as a whole grew from 8% of the total to 18% between 1890–1914.[12] This reflected a new emphasis on science as the Sorbonne, once a bastion of ecclesiastical teaching and power, continued to modernise and bring rationalism to the fore. French women were not regularly permitted to register on courses until 1902 (and they were mostly not equipped to do so as French girls' schools did not prepare their pupils for the baccalauréat until 1905). Before then however, foreign women were an increasingly visible group and by 1914 there were 1,707 foreign women registered as students in Paris. Women's preferred choice was to study arts and languages in the Faculty of Letters, followed by Medicine – the first women received ministerial permission to pursue medical studies in Paris in 1866 and the first Frenchwoman was granted a medical degree in 1875. Despite its growth and increasing importance, the Science Faculty, where mathematics was based, attracted significantly less women however.[13] In all faculties, many of these women were not pursuing degrees but registering for courses only. One woman who did this was American Ruth Gentry (1862–1917) who attended a course in mathematics for one semester at the Sorbonne in the early 1990s; she then returned to Bryn Mawr to study under Charlotte Angas Scott, receiving her doctorate there in 1896. When Marie Curie (1867–1934, then Marie Sklodowska) received her 'licence ès sciences' in 1893, she was one of only two female licence recipients in the whole university; when she received her 'licence ès mathèmatiques' a year later, she was one of five.[14] Among Marie's mathematics teachers at the Sorbonne was Henri Poincaré, one of the leading figures of the time. Despite this, there is little evidence to suggest a culture of mathematics accessible to women at Paris to rival that available to them in Germany.

Germany

For access to creative research mathematics at the end of the nine-teenth century, a preferred destination for ambitious mathematicians of both sexes was Germany. Some dozen American women mathematicians have been identified as taking the same path as Grace and travelling to Germany in search of doctoral supervision or postdoctoral study before 1913.[15] Americans were the largest contingent, but British and Russian mathematicians also took advantage of the opportunities available to women with perseverance enough to make their claim. Göttingen was a destination of particular importance and was exceptional for a number of reasons. Towards the close of century, mathematics was undergoing a transformation from traditional fields such as algebraic geometry to new, more abstract concerns such as set theory which offered immense opportunities for original research. Göttingen was leading this change and overtaking Berlin as the leading centre of mathematics. What's more, the school of mathematics (if not the University itself) had showed itself sympathetic to women and had even awarded a doctorate to a woman in 1874.

When Felix Klein arrived in 1886 to lead the department, one of the aims of this skilful politician and mathematician was to attract stu-dents of the best calibre, whatever their sex, and so ensure Göttingen's mathematical pre-eminence. (Although not a major factor, the fees paid by overseas students also contributed to the well-being of the University and women's money was just as useful as the men's.) Klein was supported in this strategy by other mathematics professors, notably the legendary David Hilbert who became a colleague in 1895. After initial struggles, Göttingen became the first German university to offer entry to mathematical women on a less piecemeal and more system-atic basis. By the early 1900s Klein had welcomed a significant number of women as researchers and, under his leadership, the University had awarded five mathematics doctorates to non-German women (Sofia Kovalevskaia 1874; Grace Chisholm 1895; Mary (May) Winston 1896; Ann Lucy Bosworth 1899; Ljubowa Sapolskaja 1902). By 1933 ninety-eight women had received mathematics doctorates from German universities, including nine foreign women.[16] This includes Emmy Noëther, generally held to be the most important woman mathematician so far, who had been awarded a doctorate from the University of Erlangen in 1907, after having been allowed to matriculate as a special exception in 1904 (and after she had studied with Klein and Hilbert at Göttingen the previous year).[17]

The admission of Grace, Winston and physicist Margaret Maltby to Göttingen in 1893 had been made tentatively as a special exception by Klein and Friedrich Althoff, a progressive at the Ministry of Culture who was in charge of higher education.[18] They planned to seek out foreign women who would return home after their studies; the access of German women to higher education was a contentious issue and they did not get the right to matriculate at German universities until well into the twentieth century. For all women, a fellowship was unthinkable and foreign women were not eligible to apply for a post at a German university after completion of their studies.[19] It was always maintained by Grace that she was the first woman to receive a doctorate in Germany in any subject, but in fact she was preceded by at least one female mathematician. Assigning absolute priority is difficult given the complexity and individuality of women's dealings with universities, however it is clear that Sofia Kovalevskaia received her doctorate from Göttingen before Grace in 1874 and that Marie Gernet was awarded hers by the University of Heidelberg the same year (1895). Kovalevskaia's experiences are instructive in illustrating the uncertainties facing mathematical women at European centres of learning.

Sofia Kovalevskaia

With universities in her homeland closed to women, Russian-born Kovalevskaia (1850–1891) travelled to Germany in 1869 along with her friend, chemist Iulia Lermontova. Both women held nihilist convictions about equality of the sexes and the power of education and were keen to forge useful, independent lives. With no formal precedents, women seeking to attend lectures were required to approach universities via individual professors on a one-to-one basis (throughout her life Kovalevskaia proved as gifted a persuader as she was a mathematician). The University of Heidelberg admitted her as an exception and she studied mathematics with two eminent professors. In class she endeavoured to keep as discreet and low profile as possible, a strategy adopted by many of the early women pioneers. Kovalevskaia's studies were not at that time leading to any formal qualification yet, after three semesters, she moved to Berlin in the hope of working with the analyst Karl Weierstrass. In response to her request he set her some problems and, when she solved them promptly and with great ingenuity, he was immediately won over and became a lifelong friend and supporter. When Berlin refused to admit her, he provided private lessons and guided her to a doctoral dissertation submitted to the University of Göttingen in 1874. Despite his sympathies however, Weierstrass was hesitant to support Kovalevskaia in gaining this formal qualification until he was persuaded that her 'arranged marriage' (which had enabled her to leave Russia) did not impose any domestic obligations on her. As will be discussed later, for a woman marital status was also a variable which affected her chances to study, especially in Germany.

Kovalevskaia's doctorate was awarded without the requirement of an oral examination, possibly due to language difficulties. Her achievement, and the novelty of Göttingen being associated with it, was remembered with fondness at the University at least until the mid-1890s: one 'old professor' kept a portrait of Kovalevskaia which he showed to Grace and other female mathematicians who were following in her wake. The example of Kovalevskaia and other nineteenth-century woman doctorates illustrates the piecemeal nature of women's admission to degrees. These pioneers, always treated as exceptions rather than precedents, were subject to the whim of individual professors and to the changing internal politics of university faculties. For example, at Heidelberg only one woman after Kovalevskaia audited mathematics lectures (American Rebecca Rice in 1873) before the Senate decreed there should be no more women;[20] this was relaxed only in the 1890s when Marie Gernet pursued doctoral study there. Weierstrass developed a personal affection for Kovalevskaia which played a part in his willingness to teach her; her thesis was sent to Göttingen because he had personal contacts there and believed the politics of the place may prove sympathetic.

Weierstrass and another of his protégés, Swedish national Gösta Mittag-Leffler, helped Kovalevskaia to gain a teaching post at the University of Stockholm in late 1883. She was now a widow and this helped smooth her path to a limited extent. Like University College London, the young University of Stockholm had been set up specifically as an alternative to the long-established and exclusive institutions at Lund and

Uppsala and had admitted women from its inception. Despite this, Kovalevskaia's recruitment caused controversy and she was initially appointed on a temporary basis without official status as an unsalaried 'privat docent' living on fees paid directly to her by her students. It was only after countering severe objections from conservative officials (concerned about her 'German mathematics' as well as her sex), and after winning the prestigious Prix Bordin of the French Academy of Sciences, that Kovalevskaia was finally appointed to a full professorship in 1889.[21]

A mathematical marriage

Despite the difficulties of finding a suitable position, after gaining her doctorate Grace was determined to follow the life of a teacher and 'celibate career woman', which is why she had refused a proposal of marriage soon after her return from Göttingen.[22] William Henry Young was a young fellow of Peterhouse College, Cambridge, who had coached Grace in mathematics briefly while her regular coach was on sabbatical. Although initially responding that she had cared for him only as a mathematician and not as a suitor, in the end Grace relented and they were married in June 1896.

The partnership seemed an ideal solution to both their needs. Pessimistic at ever finding a fellowship, Grace's correspondence reveals that she had planned that her marriage would allow her to pursue mathematical research, while her husband, with a suitably learned wife at his side, would carry on with the more mundane task of drilling students in the tricks they would need to compete successfully in the mathematics tripos. Yet by marrying and not becoming part of a community of mathematical and academic women, Grace effectively lost the female support network that had been so important while she was a student and which may have helped her to retain her mathematical identity. Female teachers had organised themselves into associations which were becoming increasingly influential as a source of funds and support for women. The forerunner of the American Association of University Women (the Association of Collegiate Alumnae) had been established in 1882 and had provided grants for women mathematicians to study in Germany. When Charlotte Angas Scott transferred from Girton to Bryn Mawr College, she became the centre of a female mathematical support network which produced six female mathematics doctorates and from which several other woman benefited, including Girton wranglers Isabel Maddison and Hilda Hudson.[23]

Grace's comparative isolation became even more significant when it became clear that her new husband was not a man to be overshadowed by his wife. Convinced that he was 'greater than the world around him', he was searching for a calling that would enable him to prove it. Grace presented it to him; in the words of their daughter, Cecily: 'It is as if she had been to him a mirror, in which he gradually saw himself and his real mind'. She goes on to describe how Grace and her husband forged a mathematical collaboration, their key aim to force recognition and a professorship for Young through publication. Cecily concludes by recalling that Grace had a special gift that she used to support her husband:

> This unique facility, which she had to the end of her life, of understanding and correctly interpreting the mind of others, enabled him to forge ahead with new ideas, building on without having continually to fill in the foundations, a job he could leave to her, and which we all know is terribly distracting and wearing in the act of creative work. When he 'laid down his staff' in 1924, it was in truth because she no longer had the stamina to help him.[24]

Whether this is an apt description of the Youngs' collaboration will be discussed in Chapter 4. Suffice to say that Grace's abandonment of her own mathematical ambitions brought her close to breakdown at New Year 1900 when she wrote, in tears, about her hopes for the future dying with the new century and the necessity of 'throwing overboard the old life' to be a wife and mother.[25]

Grace had written her despairing letter soon after the family's move to Göttingen. The couple had decided that, to be at the forefront of research mathematics, it was essential to live where all the exciting developments were taking place, and where advantage could be made from Grace's prestigious mathematical contacts from her days as a doctoral student. Göttingen was home to Grace from 1900 to 1908 and from where she managed the writing and publication of her own and her husband's papers. Young retained his teaching post at Cambridge and was therefore absent for regular, long periods of time. Grace's second sojourn in Göttingen found her in a very different situation from her previous stay; then she was a single woman of whom people held high mathematical expectations and an official member of the University. Now she was a young wife and mother (her son Francis had been born in 1897) and her relations with the University were dependent on the good will of her old professors and mathematical contacts. It is testimony to their opinion of her mathematical abilities that they welcomed her back into their community so warmly. She attended Göttingen's regular mathematical colloquium (an advanced seminar attended by professors and selected research students) as well as numerous functions at professors' homes which were part social and part mathematics. Grace's descriptions of these occasions reveal clues as to the tensions, questions of etiquette – and amusement – that could arise as to whether, and how, women could be treated as women *and* mathematicians:

> After dinner, Professor Klein was dreadfully afraid we should divide up into male and female, and he very much wanted us to talk to the men … He took my arm and carried me into the dining room to be smoked and all the men followed suit … The Professor offered me the cigars with his quaint smile.[26]

Grace also became a member of the Göttingen Mathematics Club. There is a photograph of this select group, taken in 1902, which shows Grace as the sole woman sitting at pride of place next to Felix Klein, the director of the Mathematics Faculty, who had been her PhD supervisor (Figure 2.1). The other people ranged around her are a veritable

Figure 2.1   Göttingen Mathematics Club (1902)

'who's who' of famous names from the annals of mathematics, including David Hilbert, Ernst Zermelo and Erhard Schmidt.

Conflicting roles: Woman, wife, mother – and mathematician

There is little doubt that being the only woman amongst such a select band of mathematicians appealed to Grace's vanity. In correspondence it is often recorded how she has attended a mathematical/social function and been the only 'lady' present, there 'as a mathematician and not as a woman'.[27] This distinction is a significant one because, despite the gentlemanly acceptance that Grace received from the Göttingen mathematicians, she found that sharing her time and devotion to mathematics with other duties created conflicts of self-definition within her and influenced the expectations of others. Ideals of female service were just as prevalent in Germany as they were in England and interest in 'the woman question' was especially intense around 1900 when a new German civil code was being formulated. This debate referred back to early-nineteenth-century German romanticism and stressed the importance of what was called 'the eternal-womanly': a notion of immutable femininity, predicated on woman as instinctive mother and complement to man, which was viewed as distinctly German. There was broad consensus amongst women's groups that demands should be placed within the context of what is variously described as 'intellectual', 'organised' or 'extended' motherhood. One particularly influential figure in Germany was Ellen Key whose 1898 essay, 'Misused Women's Energy', suggested that most women would not find satisfactions in their jobs equivalent to those provided by raising their own children. The side of the debate that championed equal rights with men, or offered a critique of marriage, was much less vocal in Germany.[28] A wife's support of her husband and family was expected to be her priority.

In experiencing tension between duty to marriage and motherhood or to a career, a female mathematician could be seen in much the same position as any other woman at this time who was engaged in an activity dominated by men. However, in the heady atmosphere of Göttingen, home to 'the greatest mathematical minds of the age', this stricture took on even greater urgency. Was it not imperative to support husbands who were great men doing great things for the world? Grace had surmised as much when she first arrived in Germany, remarking with approval that 'If it were not for the women, the learned men would not marry and live happily and usefully in the devotion of science'.[29] Grace was basing her comments on the wives of the professors whom she had come to know well. Felix Klein's wife Anna was a particular friend and, as the daughter of the German idealist philosopher Hegel, she may have been accustomed to providing the quiet domestic support that gifted men were thought to need. Käthe, the wife of David Hilbert, was expected to act as secretary to her husband, writing out his papers in her best handwriting and ensuring the harmonious environment conducive to his work. When Hilbert published his first paper on number theory in 1897, fellow mathematician Minkowski wrote in congratulation, also congratulating 'your wife on the good example which she has set for all mathematicians' wives'. Hilbert's biographer adds to this praise of Käthe by remarking that she never let 'tragedy hinder her husband from functioning as a scientist. Under her skillful [sic] management, the combination of fellowship, comfort and order necessary for Hilbert to work continued to be maintained ...'.[30] With these prescriptions of feminine behaviour before her, it is no wonder that Grace experienced insecurities about her role. These fears were articulated when her husband – who swung between irascibility and loving support in his relations with his wife – decided that his honour was threatened by a delay in publishing work and implied that his friends held the opinion that Grace was failing in her duty of helpmeet.[31]

There is no doubt that the Göttingen professors were unsure about how they should interact with Grace on social occasions, now that she was a wife and not a mathematics student. That her husband was not well known to them, and was absent for the majority of the time, allowed Grace greater presence as a mathematician than she might have had if her husband had been there to assume the role of intermediary. At the same time, her situation as a lone woman increased her personal relations with the female members of mathematicians' families who took her under their wing. Anxiety among male academics concerning the correct way to interact with a woman who was a mathematical peer had also been apparent during Grace's earlier stay in Göttingen, at a time when she was a postgraduate student with an official presence at the University.

Female students at Göttingen

Women students were a novelty at German universities and their arrival in the early 1890s was all too much for one young Göttingen professor who, on having to visit a new female mathematics student (Grace), had

'had an attack of politeness through nerves'.[32] May Winston recalls similar tensions in an early a letter home, describing how she was required to visit the professors prior to commencing studies and that Frau Klein had explained 'that it would be better for me, if she went with me as, in that case, I would call as a lady rather than a student and would therefore be better received'. Even so, one of the professors' wives 'did regard one as a sort of strange animal'.[33] Although Grace was impressed by the comparative freedom of life as a student at Göttingen (unlike Cambridge, the libraries, lectures and laboratories were freely available to her and she required a chaperone only when visiting one particular young, unmarried professor) female students were accorded special treatment and placed under special strictures. The women were not permitted to mix in the corridors with the male students before lectures but were to go to the professor's private room (Grace called this 'the sanctum') and he would take them in at the appropriate moment. This could be interpreted as a strategy to prevent hostility between regular students and the foreign women in their midst, more likely it was a way to preserve distinctions of sex and give the ladies due courtesy. Grace certainly saw it as the latter and often remarked in her letters home on the kindness and gentlemanly behaviour of the students and professors. Nonetheless, Grace and her two female companions sat at the back of the lecture hall (probably because they entered at the last minute) and Grace, for one, had difficulty in seeing the blackboard. However Grace did possess a key for the special mathematics 'reading room', an innovation of Klein's which was the only library of its kind in Germany. It housed a vast collection of the latest periodicals and offprints and was an important venue for mathematical interaction.[34]

Special permission had to be sought for the women to take doctoral examinations from the Ministry of Education in Berlin and they were anxious as to whether this would be forthcoming. The sensitivity of the issue was underlined by the discretion Klein asked the women to use in talking about their studies before formal permission had been granted. They were allowed to attend lectures during this time and were assured that 'the curator will look the other way when we come in'.[35] The personal recollections of both Grace and May Winston are alike in recalling the help and generosity of Felix Klein. Klein had interviewed and accepted Winston when he met her at the series of lectures that he gave at Evanston during the 1893 Congress of Mathematics in Chicago (Christine Ladd-Franklin was also there and offered Winston a year's scholarship to enable her to take up the place). Once in Göttingen, Klein had visited Winston's lodgings in person to help her with her application to the Berlin Ministry charged with giving permission for women to sit the doctoral examination.[36] Klein was wise to pursue his plans with caution; male academics were not used to having women in their midst. Limited experiments with female 'auditors' at lectures (auditors were permitted by special permission, not by right, and denied student status) had been made in Germany from the 1870s onwards. The majority of these women were foreign, predominantly British, Russian and American. It was six years later, in the summer term of 1900, that German women were for the first time entitled to become fully-registered students and sit examinations. Heidelberg University was the first to yield; Prussia did not open its doors to women on a formal basis until 1908. One of the reasons for Germany's relative tardiness in providing higher education for women was that, with a well-established system of state-controlled education, there was no tradition of private philanthropy to provide impetus for new initiatives. Private colleges did not prosper to the extent that they did in Britain and America.[37]

Debate concerning women's potential for higher education

Germany and Britain were alike however in respect of the war of words waged over women's capacities for higher education. Around turn-of-century women's access to university was a subject of debate made especially intense by the belief that the Universities, since unification in 1871, were a crucial factor in the formation of a strong German identity. The threat of women harming their status was underlined by the gendered concept of *Bildung* (education as a way of developing the self) which placed men, especially University men, as the natural agents of cultural production. For a woman to assume this role was a contradiction of her femininity, especially if she opted for a traditionally 'male' sphere such as mathematics. Katharina Rowold has uncovered how Sofia Kovalevskaia's early death was explained by contemporaries with reference to the strain put on her psychological well-being by her dedication to mathematics and the life of the intellect.[38]

In contrast to the mathematics faculty, other departments and the administrative authorities at Göttingen were strongly against the admission of women. Max Runge, Professor of Gynaecology, argued that women's physiology made them weak and incapable of academic study, and that their whole organism had reached a less advanced state of evolution. Göttingen's highest official was also against Klein's plans to recruit female doctoral candidates and accused him of proposing 'a notion worse than social democracy, which only seeks to abolish the difference in possessions. You want to abolish the difference between the sexes'.[39] In fact, that was the last thing that Klein wanted to do. He argued for English-style single sex colleges for women and believed that medicine, law and theology were inappropriate for the female mind to study. Unlike mathematics, these subjects could introduce women to unwholesome ideas, expose them to unseemly public display and take them into areas from which God had ordained they should be absent. Grace was particularly nervous of telling Felix Klein that she intended to undertake part-time medical studies at Göttingen, no doubt fearing his displeasure. Grace gained a Medical Students' Registration Certificate on October 25th 1900, having decided to embark on medical training as a 'fall back' career. (Grace completed her formal medical studies but, as mathematics came to dominate her life, did not continue to complete the final, hospital-based training.) Grace was accepted as a student by the medical faculty as a special case and it was made clear that only experienced married women would be considered, a viewpoint with which Grace strongly agreed.

The idea that women could not produce high quality academic studies without relinquishing their femininity or health was a frequently-voiced opinion in Germany. Again, mathematics and the abstract sciences were singled out as being particularly at odds with female nature. The neurologist Paul Mobius in *On aptitude for mathematics* (1900) warned that 'a mathematical woman is contrary to nature, a kind of hybrid'[40] and Max Planck, Director of Theoretical Physics at Berlin University, similarly cautioned against eroding natural sex difference. It could 'not be stressed enough that Nature herself assigned to women the role of mother and housewife … to ignore natural laws is to invite great damage which will in this case be inflicted upon coming generations'.[41] Planck was alluding to fears during the 1890s, paralleled in England, surrounding Germany's perceived declining birth rate and high number of infant mortalities. The finger of blame was pointed at women for threatening national wellbeing by neglecting their role as mothers or losing their capacity for child bearing through inappropriate (masculine) intellectual activity. That Grace was influenced by these arguments can be evidenced from her endorsement of Max Runge's pamphlets, and by her support for Francis Galton's views on the need to encourage childbirth amongst the educated middle classes.[42]

Despite Grace's favourable accounts of her experiences at Göttingen, there is evidence that there was some antagonism towards foreign women who were benefiting from an education unavailable to German women. Physicist Margaret Maltby, who studied alongside Grace for a doctorate at Göttingen, gave an address in 1896 in which she reported that 'Many German women, and men, too, feel strongly the injustice of an attempt on the part of foreign women to avail themselves of certain university courses merely for the purpose of general culture'. She went on to argue that far more than just an individual woman's wishes needed to be considered: 'Let her admission come about in a way to command the respect of professors and students'.[43] This concern is one of the reasons why Felix Klein was so adamant that he would only accept women students if they were proven 'capable of contributing'. Grace was quick to endorse this standpoint and stress that she did not want 'the universities to become the happy hunting ground of the mere knowledge desirer'.[44]

Felix Klein had been thwarted in an attempt to have another female mathematician admitted to his department in 1891 and the admission of Grace and her two female colleagues may have been a first victory for him in a long-running battle. As discussed above, American Christine Ladd-Franklin had completed all the requirements for a doctorate at Johns Hopkins University only to be denied her degree because of her sex. She had travelled to Göttingen with her husband in the hope of participating in university courses alongside him. When she was refused, Klein took up her case and argued for her to be accepted as a regularly-matriculated student. Klein was defeated in this and Ladd-Franklin was permitted entry as an 'auditor' only without student status.[45] Klein's later initiatives in gaining access for women were taken up quickly and enthusiastically by what became a network of supportive and generous women mathematicians. When May Winston joined Grace at Göttingen, it was with the financial support of Ladd-Franklin who contributed $500 towards her costs after being made aware of Winston's financial worries by Charlotte Angas Scott. Angas Scott also sent Isabel Maddison, Grace's close friend and fellow student at Girton, to spend a year with Felix Klein on a Bryn Mawr scholarship between 1894–5. Towards the end of the century, a small number of women in Göttingen mathematics lectures was not unusual. Klein was not disappointed by his 'experiment': he wrote of Grace and her two American colleagues:

> We have had the most positive experiences: our three women are not only exceptionally diligent and conscientious, but their accomplishments are in no way inferior to those of our best students; indeed, to some extent they serve as a model for them.[46]

Motherhood, eugenics, Nietzsche and mathematics

A milestone in Grace's journey from a young woman who intended 'to lead a different life' to one who subscribed, in theory at least, to Victorian ideas of female service, was her writing of a private book for girls and their guardians entitled 'Mother Nature's Girl'. Here Grace argues that the first duty of a girl was to make her body a fit one for her to perform her duties as wife and mother – arduous duties which were no less valuable to the State than those of a soldier. The first duty of a boy was to serve his country.[47] Although Grace came to embrace ideals of 'the eternal motherly' in her writings, she did not pursue them in her daily life. Her unmarried sisters-in-law carried out the 'mothering' of her children, aided by a succession of hired 'girls', while she devoted her time to mathematics. This division between theory and practice is testimony to the conflicting influences that Grace was attempting to manage. On the one hand she loved mathematics and felt obliged to contribute something meaningful to it, on the other she acquiesced to the importance of duty to husband and family. Grace's views on motherhood complemented a growing interest in the emerging eugenic movement first articulated while she was a student at Girton. Her mother had been an admirer of Francis Galton and had contributed family data to him when he was collecting family statistics on inheritance and intelligence in the mid-1880s. Galton defined eugenics as a science concerned with understanding and controlling the influences that improve the qualities of a race, particularly heredity. In 1909 Grace resumed the family's contact by writing to express an interest in joining his new society. Galton had just published a paper in the *Westminster Gazette* in which he drew attention to the new Eugenics Education Society and this prompted Grace's renewed interest. In her letter Grace wrote that it was 'evident that the aims you advocate are precisely those that my husband and I have at heart and which we have been both practically and theoretically working to advance'. This was a reference to her six children whom she described as especially talented and 'a test case for many questions in heredity, both for mental and for physical peculiarities'.[48]

According to Greta Jones, the Eugenics Education Society was an essentially middle-class group which tended to emphasise intellectual achievement and the inherited aristocracy of talent.[49] This description is an apt one of Grace and her concerns. Implicit in eugenics was the belief that the most important human characteristic was mental ability and that this was inherited through the generations. This idea, which implied a *natural* (not social) elite of gifted intellects, sat well with a pure, highly abstract mathematics which was believed to be the creation of the 'greatest minds'.[50] In Germany as in England, eugenic ideas were beginning to gain a foothold as the country went through similar processes of industrialisation, declining birth rate and social change. Concerns for racial hygiene had found expression in a movement active since the 1890s and a Racial Hygiene Society had been established in Berlin in 1905. As well as concerns over diseases and social problems such as alcoholism, women giving birth to fewer than two children were also targeted as exhibiting deviant behaviour.[51]

Grace's admiration for Galton was complemented by her growing espousal of the language and ideas of German thinker Friedrich Nietzsche, quotations from whose work appear in Grace's personal notebooks. Although Nietzsche's last book had been published in 1889, it was not until the mid-1890s that he became a figure of influence whose ideas were pressed into service by those involved in intellectual, social and political activities in Germany. Although Nietzsche's work can be obscure and open to interpretation, his principal ideas centre on contempt for Christian 'slave morality' (with its compassion for the weak) and a glorification of the 'superman' who has the will to dominate and is above ordinary morality. This elite *Übermensch* is part of a 'master class' whom the rest of humanity are there to serve. The *Übermensch* is always characterised as male; for Nietzsche, women lacked the 'will to power' that superior human beings possess and so were destined for servitude: 'Women want to serve and find their happiness in this'.[52] The term 'eternal-womanly' is used by him to emphasise woman's moral duty towards man as helpmeet.[53] Nietzsche constantly reiterates that there are higher human beings who are more valuable than the mass and for whose sake the mass exists and must be sacrificed. These sentiments are reflected exactly in a poem written for Grace's son Frank (Francis) who was killed in 1917 during World War One. Using Nietzschean terminology, Frank is called 'Superman', 'the Great Exception', the 'Arab steed' in the presence of 'carthorses'. Adding Galton's ideas to Nietzsche's philosophy, the poem argues that the 'Superman' can be identified though the use of the 'new intellectual tests'.[54]

Of course, this poem was written in the despair of death, but ideas about the superiority of the ideal male intellect were an ever increasing preoccupation of Grace's since her days at Girton. According to Grace's notes at least, Young (who in her autobiographical jottings she calls, with obvious connotations, 'Mr. King') had proclaimed himself 'greater than the world around him', citing this as the reason why his abilities went unrecognised.[55] In the early days of their marriage at least, Grace was preoccupied with showing to the world that her husband was no ordinary man, a preoccupation made more urgent by her family's ambivalence to him and his failure to acquire a suitable post. She compares him to Socrates and Browning in her letters and writes that she finds 'my greatest comfort and joy in reading about really great men'.[56] Grace's glorification of the male mind can only have been confirmed by her experiences in Göttingen.

Mathematics, 'great minds' and masculinity

Despite the sympathy to women students shown by the mathematics department, Göttingen's masculine rituals, and the style of mathematics pursued there, were all informed by a glorification of the superior, masculine mind. Unlike mathematics at Cambridge University, which by the 1880s was just beginning to catch up with Continental developments in analysis, Göttingen privileged research and promoted creative mathematics. The University's celebrated professors – Felix Klein, David Hilbert, Hermann Minkowski and others – established schools around them which attracted 'disciples' from all over the world. In this, Göttingen became a model for later Cambridge and the new American universities. Within this structure, admiration of individual intellects may easily turn into myth-creating glorification. Contemporary accounts emphasise the awe with which these individuals were viewed, an interpretation repeated by later accounts that present these 'mathematical heroes' within the standard narrative of eccentric genius.[57] Klein is described variously as 'regal', 'kingly', 'Olympian' and even the 'divine Felix'.[58] After their first meeting, Grace gushed that he was born to lead and that 'Cambridge has nothing to compare with him, both as mathematician and as man'.[59] Apocryphal stories surrounded him such as if dining at Klein's house a student was sometimes so awed by his host that he stood up when he was asked a question.[60] Hilbert was also the subject of mythologising anecdotes. These centred not only on the abstraction of his mathematics, but on his penchant for shabby clothes, inveterate womanising, and his infamous mathematical walks during which students followed him around Göttingen while engaging in mathematical brain storming. The celebrity of these 'great men' reached its apotheosis in 1912 when a series of postcard portraits of Göttingen mathematicians went on sale in the town (Figure 2.2). Although the male mathematics professors and students behaved with gentlemanly politeness to their female

Figure 2.2 Postcard of David Hilbert (1912)

peers, there is evidence that Göttingen's women mathematicians relied on each other for mathematical and social communication, creating their own parallel networks. Both Grace and May Winston recall the ways in which all the women relied on each other for support and

Winston was particularly grateful to Grace for providing 'an opportunity which I have never enjoyed before – to work with someone who is my equal – or nearly so – for I do not consider her *very* much ahead of me'.[61]

Mathematics dominated Göttingen and the students, too, enjoyed special status. One journalist recalled the lordly young men who wore 'their caps visored in the bright colours of duelling fraternities, their faces usually swathed in bandages.' They left behind them 'a nauseating odour of iodoform which penetrates everywhere in Göttingen'.[62] In common with other German universities, duelling was a tradition at Göttingen. As duelling scars were regarded as badges of courage and honour, cuts on the face were left open to heal so that they would leave telltale marks. Grace echoes this view of Göttingen's duelling culture, where students fought duels on a Saturday night before an audience, in her fictionalised autobiographical writings. Here she has the daughter of a 'delightfully patriarchal' professor's family remark that:

> I never look at a student who has not yet got *Schnitte* (cuts on face).
> If you look at the students on the Weender Strasse (Straße), when
> between 12 and 1 they parade up and down, especially on a Sunday,
> you will see them all bandaged up and the whole place smells of
> carbolic![63]

By the 1890s duelling was frowned upon by the authorities yet it was still a favoured activity amongst undergraduates; it was even endorsed by the Emperor in a speech at Bonn University in 1891.[64] Most common was the 'assigned' duel (for sport rather than to resolve a dispute) for which students dressed in a special padded leather outfit including trousers, gloves, silk around the neck and wrists, and a felt hat. These duels were widely practised and university officials tended to turn a blind eye to the activity.[65] Duelling can be interpreted as the German equivalent to the bodily training and exercise that accompanied preparation for the competitive mathematics tripos at Cambridge. The structure of the duel – one man's physical strength and cunning pitted against another's – was the model for interpreting the activity engaged in by mathematicians. In Göttingen, individual minds were seen as the producers of new mathematical knowledge. The professors were engaged in 'intellectual duels', using their brilliantly-penetrating minds in place of swords, and the victors were those who published first and received the credit, often attaching their name to a theorem that would be remembered by generations of mathematicians to come.[66] The professors bore their scars too. Felix Klein suffered a mental breakdown said to be caused by rivalry with Henri Poincaré who was developing similar ideas at the same time. The structure of the Göttingen mathematics department reflected the idea of competition too. Access to seminars was restricted to top students who had 'won their spurs' in what has been called a 'highly competitive and unashamedly elitist approach to mathematical education'.[67] (Both Grace and May Winston wrote of their excitement and nervousness at being required to give a paper in front of the mathematics professors at Klein's seminars.) Severe competition was also a feature of meetings of the Göttingen Mathematical Society where 'the atmosphere (was) anything but relaxed'.[68] But for a woman to be a competitor, especially alongside men, was problematic. Men were prepared for competition, but ideals of femininity as complement and helpmeet to man dictated that women (middle-class women at least) should live private lives and not engage in public debate with men. Although these ideals were not rigid but evolving as women negotiated demands for higher education and the vote around them, Grace was not alone in adhering, in theory at least, to such codes of female behaviour. Much opposition from men and women to the suffrage movement pivoted on the unseemliness of women speaking on public platforms and acting in an altogether 'unladylike' manner. Grace expressed similar sentiments when she explained her anti-suffrage stance on a dislike of 'the argumentum ad hominem' of the suffragists which 'degraded' the whole subject.[69] Grace's husband engaged in a bad-tempered dispute (or duel) over priority with Schönflies which was played out in mathematics journals during the mid-1900s;[70] it would have been difficult for Grace, as a woman, to defend her position so robustly and acrimoniously as her husband was able to do. Grace's decision to allow her husband to compete in the mathematical world, with her 'back room' support, can be seen as a compromise that (in part) reconciled her mathematical ambitions with this sense of feminine propriety.

Göttingen's standing in the mathematical world at the turn of the nineteenth century was epitomised in its reputation as 'the Shrine of Pure Thought';[71] at that time Göttingen-style abstraction was the dynamic behind nothing less than a reformulation of the foundations of mathematical enquiry. For Mary L. Cartwright, Cambridge scholar, Mistress of Girton 1949–1968 and, in 1947, the first woman mathematician to be elected a fellow of the Royal Society of London, Grace's work at Göttingen had 'an enormous influence on the development of Cambridge mathematics'.[72] Grace and her husband were at the forefront of the new mathematical analysis, most of their work being developments of Georg Cantor's theory of sets. Felix Klein made a habit of recommending particular areas that he felt were ripe for development and that his students could take further; this was the subject that he had recommended to Grace and, through her, to her husband. Cantor's work used powerful ideas, such as irrational numbers, complex numbers and differing powers of infinity, which enabled equations to be solved and mathematics to progress. It had opened up entirely new areas for mathematicians to work in, almost an early twentieth century job creation scheme. This is part of what David Hilbert was implying when, in response to criticism that these concepts were 'unreal' and therefore should not be a part of mathematics, he famously said that mathematicians must not allow themselves to be 'thrown out of paradise'.

Although this new analysis had important applications to physics, at the research level it was far removed from worldly problems. Exponents of this new mathematics believed that existing methods had come to a dead end. It was as if they were using a ruler but the thing that they needed to measure went not along the ruler but behind it, above and below it, and between its markings. A pair of variables represents a point in the plane and a triple represents a point in three dimensional space, but as the number of variables the analysts worked with increased so they found it necessary to posit spaces of a higher dimension. In 1905 David Hilbert produced his famous theory of infinitely-many variables known as 'Hilbert Space Theory'. Grace and William Young presented their 1906 book, *The Theory of Sets of Points*, as 'the first attempt at a systematic exposition' of this new analysis and an 'attempt to further the frontier of existing knowledge'.[73] Cantor's new analysis provoked objections that the concepts it introduced (such as multidimensional space and many infinities, some greater than each other) were not grounded in reality. One of Germany's leading mathematicians, Leopold Kronecker (1823–1891), fought unsuccessfully against Cantor's work being published. Kronecker objected strongly that, with this new type of mathematics, the idea of mathematical truth involving some kind of correspondence to the real world was entirely jettisoned. Instead, truth lay in internal coherence and absence of contradiction; assumptions could be based on anything (not just observable facts) and they usually emerged from the creative minds of individual mathematicians. No wonder that these brilliant men seemingly creating mathematics anew, were revered and mythologised.

Pure mathematics, mathematical physics and engineering

The journey of pure mathematics into abstraction, and adoration of the intellects that produced it, can be seen in part as a response to the growth in status and popularity of mathematical physics and engineering. In Germany, as in England, there was a growing opposition between pure and more practical mathematics, the pure mathematicians becoming increasingly elitist as they produced mathematics that only a small group of experts could read, let alone understand. At the same time the new physical sciences were engaged in expanding their sphere of influence and producing technology that could be seen to be changing the world. As a result, the more practical sciences were becoming popular subjects at university, challenging the historic prestige attached to purer mathematics. In Germany, tensions between these two specialising disciplines led to a revolt amongst mathematical and engineering teachers who wanted to reform mathematical education to cater more specifically to practical needs. Where the new engineers were democratising their discipline with technical institutes to train large numbers, the pure mathematicians were shunning any reference to the 'real world' as a contamination and emphasising the ability of only special individuals to further mathematics. Grace constantly put forward the view that what was important was not the success or utility of mathematics but 'the change in the mental point of view of intellectuals'.[74] Furthermore, it was 'not the usefulness of mathematics that constitutes its claim to be a form of expression for the beautiful' and any application to the real world was 'merely coincidental'.[75] Grace concludes by likening mathematics to Art, although arguing that mathematics is more difficult and more 'occult'.

This emphasis on the transcendence of pure mathematics is echoed by Georg Cantor who maintained that his theory of sets belonged 'entirely to metaphysics'.[76] Eleanor Sidgwick betrays similar moral concerns when she states of mathematics that a subject not studied for its own sake, but because of its usefulness for something else, 'is almost degraded in the process'.[77] G.H. Hardy, one of the first mathematicians to take up the new analysis in England, echoes these sentiments in *A Mathematician's Apology*, his account of mathematics which was first published in 1940. Here he argues that mathematics cannot be justified for its 'crude achievements', that 'Real' mathematics was 'almost wholly useless', and, furthermore, that this very remoteness from ordinary activities kept it 'gentle and clean'.[78] Hardy also likened mathematics to Art, adding that the mathematician's 'beautiful and harmonious patterns' are more permanent because they 'are made with ideas' and there can be 'no permanent place in the world for ugly mathematics'.[79]

This was a philosophy entirely compatible with Nietzschean ideas about elite individuals and in keeping with the adulation of intellect becoming visible at Göttingen. Pure mathematics was the preserve of a (male) aristocracy of talent and the inexact, utility-driven maths of the growing band of engineers and technicians was as anathema as the new movement for socialism. This view retained echoes of the ideals of a liberal education. This tradition, encapsulated in the learning of dead classical languages, was in part based on the idea that knowledge, when applied to material ends, was an impure subject unworthy of the attentions of the intellectual elite. It was also a view coloured by class: the engineers and practical scientists engaged in providing education and technology in England were concerned to assert their 'professionalism' as a way to achieve parity with more established, less 'hands-on' disciplines, as will be illustrated in the succeeding chapter. In Germany too, the issue of the relationship between pure and applied mathematics, and between its practitioners, was a complex one.

In Cynthia Cockburn's classic materialist view of history, the makers of tools, from the early days of hand-held implements to today's computers, achieved status and power from their skill. The making of tools was confined to men, as toolmakers acquire power and authority in as far as people are rendered dependent on them.[80] In a move that echoes this analytical model, in the decades surrounding 1900 pure mathematicians identified a role for themselves in creating tools for the more practical mathematicians. As Klein explained to an audience of engineers, 'Mathematics' business was to formulate the fundamentals'; from where hypotheses originated, or whether they were based on observable fact, remained of no consequence. Mathematics was 'not responsible if the consequences of deductions do not correspond with reality'.[81] Despite this remark, Klein was acutely aware of the dangers in Germany of abstract-oriented, pure mathematics and mathematical education becoming isolated from the practical. To ensure this did not happen, he sought always to retain an holistic view of mathematics, investigated ways to apply his abstract research to practical ends, and in 1898 initiated a new university curriculum which introduced the teaching of applied mathematics for the first time.[82]

In contrast to Klein, for Grace and her husband the hierarchical relationship of the pure and the applied was often centre-stage; they introduced their book on analysis with a prediction that 'the near future will see a marked influence exerted by our theory on the language and conceptions of Applied Mathematics and Physics'.[83] This sentiment is echoed by Grace when she asserts that 'In the Dark Ages Theology was the Queen of Science. In the time to come she will yield her place to Mathematics of the purest and most abstract kind'.[84] The intrinsic superiority felt by the pure mathematicians is summed up by a contemporary anecdote that David Hilbert maintained that he had taught himself physics because 'Physics is too hard for physicists'.[85] This sentiment would have been appreciated by G.H. Hardy who wrote that applied mathematics 'is dull and for dull intellects'.[86]

Genius, gender and mathematics

It is not coincidental that both Grace and G.H. Hardy liken the pure mathematician to the artist. The romantic idea of genius had particular applications to Art. Although romantic philosophy had its roots in an early-nineteenth-century reaction to rational materialism, Victorian thinkers raised the ideal of the male genius almost to a 'cult'.[87] Here the genius is born with talent; endowed by Nature with a rare gift as opposed to having learned a skill. The surrounding narrative presents this exceptional individual (artist, poet, mathematician, composer) as an isolated figure who uses his creativity by force of spirit alone. This was a potent scenario in Germany in particular, drawing on a tradition of German idealism through Kant, Fichte, Hegel and Nietzsche, which stood in contrast to British inductive pragmatism. The latter, through the work of philosophers such as John Locke and David Hume, privileged empiricism, used inductive reasoning, and argued that all knowledge was ultimately derived from experience. In contrast, German idealism, whose 'golden age' was the 1770s to 1840s, had connections with Romanticism and was concerned with notions of 'pure thought'. Abstraction and 'purity' (which was both the process and the product of the new mathematics) can be interpreted as a counterpart to the pure spirit which was held to work through the genius as Nature used him as its instrument. No wonder pure mathematicians were concerned with the development of individual minds and not with the bodily world.

This dualism of mind and body, which rejects the world as corrupt, can be traced back to Descartes. At the end of the nineteenth century, this rejection can also be seen as a reaction to Darwin's materialist placing of man in the world alongside the animals, and a part of what is often called the 'crisis of the intelligentsia'. For example, Hardy's privileging of the 'pure' and notions of genius can be interpreted as an attempt to retain some idea of transcendence after he had lost his faith and become an 'ordinary educated unbeliever'.[88] The work of genius transcended normal confines and the more abstract a piece of mathematics, the more moral credibility it possessed. Genius had no truck with the bodily, which is one of the reasons why it was argued that women, rooted in the material world of reproduction, cannot possess it.

Although attempts were made by some feminists, in Germany and in England, to claim genius for the female sex, it remained a concept with strong masculine connotations. Christine Battersby has written a comprehensive study of the history of gender and genius, tracing the concept from its prehistoric roots and demonstrating its connection with male sexuality. Battersby describes how 'genius' and 'woman' became conflicted ideas towards the end of the eighteenth century. At that time, a reaction against Enlightenment 'reason' resulted in a new discourse of creativity (centring on emotion, sensitivity and 'feeling') and this was then mapped-on to existing ideas about masculine virility:

> This rhetoric praised 'feminine' qualities in male creators but claimed that females could not – should not – create ... The psychology of a woman was used as a foil to genius: to show what merely apes genius. Biological femaleness mimics the psychological femininity of the true genius. Romanticism, which started out by opening a window of opportunity for creative women, developed a phraseology of cultural apartheid ... The genius was a male – full of 'virile' energy – who *transcended* his biology: if the male genius was 'feminine' this merely proved his cultural superiority. Creativity was displaced *male* procreativity: male sexuality made sublime.[89]

This gendering of genius continued and, around 1900, the idea of female genius was marginalised in the works of Galton and sexologist Havelock Ellis. Galton claimed that genius was inherited down the male line, adding that, on the rare occasion that women did possess genius, they would become so masculinised that they would be unmarriageable and so would be unable to pass on their ability. For Ellis, male genius was closely linked to men's sexual energy or 'vital force' and was exhibited in intellectual and creative spheres; women's 'vital force' was used up in reproductive duties, so any feminine genius was manifested through 'love'.[90] The message was clear: the best a woman could hope for was to give birth to genius, not to become one herself.

Control of bodily impulses and sexual energy as a condition of authentic manliness has been identified as another key notion at the end of the nineteenth century. At this time, chastity and men's literature presented a struggle between 'mind' and 'body' and 'reason' and 'passion' as a key part of male experience. A distinction was made between 'manly' and 'male'; a man achieved manly status through self-mastery of his body and his sexual impulses.[91] (It was this kind of thinking that led to the rigorous physical training that accompanied competition in the Cambridge mathematics tripos.) In this way, male elites conceptualised masculinity as rational, thereby constructing femininity, by contrast, as emotional. The Göttingen analysts, concerned as they were with pure thought, were operating within both the narratives of disembodied genius and of manliness. It is notions such as these that underpin Grace's advice to women (specifically her daughter) thinking of entering university: 'The sensual life of a man, once awakened, eats up his powers' she warns, therefore 'it should not begin until the training of his career is over'. Furthermore, inflammatory women in academia could 'cause infinite harm to her comrades, sapping their strength and ruining their prospects. She herself is the sufferer, she herself who, whatever her personal ambitions, really longs for a superior male mind'.[92] It is significant that Grace used the terminology 'male mind' and not 'man'. Engineering was comprised of men concerned with the material and bodily; pure mathematicians were 'superior', concerned with only the abstract and the mind. Furthermore, great mathematicians must be enabled to use their creativity and virility for mathematics, not tempted to dissipate their powers of genius in early relations with women.

For the pioneers who had brought an earlier version of continental analytics to Cambridge in the early nineteenth century, genius was a redundant concept. As outlined in the preceding chapter, Babbage and Herschel rejected established authority in favour of learned technique and characterised their mathematics as a way to modernise and democratise the process of thought.[93] The new analysis at the rise of the twentieth century was very different. The earlier mathematicians had wanted to get rid of the subject and concentrate on objectivity. For Grace and her peers, the subject and the quality of his individual mind was crucial. The earlier analysts wanted to apply their calculations to the world; the new analysts judged their success by how far removed from the world they could make their mathematics. Old style analytics was a technique that the rising middle classes could learn and use as a weapon in their assault on aristocratic learning and privilege; new style analysis was an activity for a gifted elite. Both Grace and the Cambridge reformers criticised the mathematics tripos for its function as a treadmill for technique and quantity at the expense of creativity; like Cambridge philosopher William Whewell before them, they imported an ideology of romanticism and genius which privileged pure mathematics and the mathematicians who produced it. David Bloor has argued compellingly that even deductive, formal systems of knowledge are socially produced as their underlying premises, principles and assumptions are subject to negotiation and social acceptance.[94] New ideas deriving from Cantor's theory of sets, notions such as multidimensional space, many infinities and irrational numbers were controversial at first and split the mathematical community. There was no certainty that mathematicians would not be thrown out of Hilbert's 'paradise'. However, Grace and her peers were amenable to a system of mathematics which was unsullied by the 'real world', rested its moral credibility on abstraction, referred back to a tradition of neohumanist ideas, and preserved the status of an elite of intellectuals in the face of a challenge from the new, practical disciplines.

The language of genius

The abstraction and moral integrity of this style of mathematics was increasingly conveyed by a re-emphasis of language that was highly feminised and romantic. Proofs were 'elegant'; theorems 'beautiful'. The aesthetics of theorems – simplicity, consistency, symmetry – became viewed as an element with which to judge their success. Bertrand Russell, who concerned himself with the logical philosophy underpinning analysis, believed that 'Mathematics, rightly viewed, possesses not only truth but supreme beauty ... sublimely pure'.[95] This feminised language distinguished pure mathematics from the applied and helped women such as Grace to feel comfortable within the discipline. However, key to this language was its correspondence to the same androcentric idea of romantic genius described above; an idea that found its expression in feminised language, but its effect in excluding women.

As previously outlined, by the end of the eighteenth century, the idea of genius was understood as intimately connected with the creativity of male sexual energy; in contrast femininity was constructed as at core emotional, sensitive and intuitive. These latter qualities conflicted femininity with rationality, an attribute coded inherently male. However, one of the defining features of the genius was that he was sensitive and guided by instinct like a women, but also had the body and rationality of a man. In this way, the genius was a 'third sex';[96] an embodied male able to access and use 'feminine' attributes in a way beyond any woman's reach.

By using feminised language, pure mathematics was able to reconcile abstract, 'masculine' rationality with 'feminine' intuition and instinct, a dichotomy that resulted in genius. In Germany, this association was strengthened by a particular construction of elite masculinity that presented the ideal, 'whole man' as a combination of male and female ideals – someone 'who combines all of the qualities that are separated in the polarised gender model'.[97] It is this view that Grace may have been alluding to when she wrote of yearning for the support of 'the complete man'. But not content with being credited with genius, the great mathematicians at 'the Shrine of Pure Thought' were deified too. One student of 'the Divine Felix' maintained that even after he came to know him well, he still felt the distance between himself and Klein 'as between a mortal and a god'.[98] On her arrival in Göttingen, Grace quickly acquired similar terminology, remarking on 'epistles' that she had received from Klein and calling his rooms 'the sanctum'. The notion of God as a divine mathematician has had currency since the days of Pythagoras; for Margaret Wertheim this is just a short step away from the idea of mathematicians as divine, and the reason why women have been excluded from the 'priestly' practices of mathematics and physics.[99] Certainly, there were few models of female transcendence or genius that Grace found to negotiate an identity around. Instead, surrounded by prescriptions that served to limit female ambition and opportunity, overwhelmed by romantic notions of genius, and practising a pure mathematics that privileged the male intellect, she decided to transfer her mathematical ambitions onto her husband.

Conclusion

When Grace attended a mathematical dinner held in honour of Felix Klein, her lone female presence was explained by the fact that she was 'there as a mathematician, not as a woman'. The tension between these two terms was still acute enough to require amplification. In this world of pure mathematics, where the worth of a particular piece of mathematics was based not on its success or utility but on its beauty and internal consistency, and where what constituted a 'proof' was open to debate,[100] it was essential to situate yourself as a mathematician of high ability and gain the trust of your audience that this was so. Grace found it increasingly awkward to characterise herself as such, or to contest the undercurrents in the Göttingen mathematical community that made it difficult to see how she, as a woman, could – or should – aspire to emulate figures such as Hilbert and Klein. Overwhelmed by romantic notions of genius, she seemed to lose confidence. Grace continued to research mathematics, was in constant correspondence with leading mathematicians, and remained highly respected within the mathematical community. However, she severely compromised her personal ambitions and, instead, used her mathematical connections and skill to support her husband. The couple's collaborative practices will be explored in Chapter 4, suffice to say that Grace published only four papers under her name alone (including her PhD dissertation) up to 1915; she resumed only after 1914 when Young was finally awarded a professorship. (Grace's later work was influential in the field of differential calculus and one of her papers won the Gamble Prize for Mathematics, awarded to Girton alumnae, in 1915.) Despite this, Grace's exposure to the romantic culture and ideology of 'the shrine of pure thought' led her to question her identity as a mathematician and to elevate her husband (not herself) to the same level as the great Göttingen mathematicians whom she revered: a small and elite club that was open, she came to believe, only to men.

3
Professional or Pedestal?
Hertha Ayrton, a Woman
among the Engineers

In 1898, the Electrical Trades Union (ETU) sought to symbolise its work and purpose with a new, specially-commissioned banner bearing the words 'Light and Liberty' (Figure 3.1). This classically-inspired image invites a number of interpretations and allusions, each with gendered meanings. The banner is dominated by the central figure of a glorious 'angel of light', a female personification of both electricity and the union, who presides on a pedestal while spreading her wings above six male figures.[1] The latter, unlike the abstract 'angel', are representations of real, individual men made recognisable by being clothed in the identifiable garb and accoutrements of their differing trades. These workmen reach up to the female figure to acquire and master the wonderful power of electricity which they then control and fashion into technology, while the 'angel' simultaneously seems to protect them and offer up her fabulous, natural power.

In the final decades of the nineteenth century, as electricity and consumer culture developed hand-in-hand, electricity could be represented, not least by companies marketing their new electrical devices, as an almost magical, invisible force that could be conjured and used for a myriad of domestic, medical and other purposes. From the late-1880s electric lighting was increasingly replacing gas in the homes of the well-heeled middle and upper classes, while medical and other devices (such as the galvanic equipment that Grace Chisholm's invalid sister used to treat muscles wasted by polio) were promising the chance of a better, more comfortable life. In this context, the use of the female form in industrial images reflected an acknowledgement that women were users and consumers of new electrical technology, even if they were not engineers of it.

Figure 3.1   The Electrical Trades Union Banner (1898) by permission of People's History Museum

But these feminine representations served a purpose beyond just the recruitment of women customers: 'The goddess archetype helped lend an aura of dignity, legitimacy, and stability to a world of rapid mechanization and technological change'.[2] The electricity industry in particular used classically-draped and posed 'electric eves' as emblems of the electric age.[3] These mythic figures, whose bodies were often transformed magically through the use of electricity, were used to decorate technological products, on technical literature, and as huge sculptures at exhibitions and trade fairs.[4] It is clear that the use of the classical feminine archetype on the ETU banner was designed to surround this young, working class union with an aura of moral worth; it also suggested status and professionalism. The ETU had been formed in 1868, after the Amalgamated Society of Engineers (formed 1852) had refused membership to electricians. The ETU's new banner affirmed its parity with the engineers and, in a society that traditionally privileged 'intellectual' over 'practical' work, represented the electrical trades as more than just 'working class' labour. Around the same time, the Institution of Electrical Engineers (IEE) had used a similar strategy to convey messages of moral integrity and to inspire unity among the 'practical' and 'professional' men within its membership. On the centenary of his birth, the telegraphic and electrical engineers of the IEE invoked the figure of Michael Faraday as 'founding father' of their specialised profession, creating a mythology around him and using his image on their first institutional seal.[5]

While the new ETU image reflected metaphors of science that had long personified the world and her forces as female; it also acknowledged that the technical business of engineering this new force of electricity, and using it to bring important new benefits to the service of mankind, was a wholly male enterprise. Middle-class women such as Hertha Ayrton are subconsciously invited to identify with the 'goddess' figure (typically the only female representation within such images) rather than with the men who mediate her power. In this way, the banner reflects contemporary society's expectations of women and the 'decorative', passive roles considered appropriate for them to assume. The gendered messages conveyed by the ETU banner are indicative of the representations and experiences of Hertha as she attempted to participate as a peer, from the mid-1880s until her death in 1923, in the worlds of electrical engineering and experimental physics. By following Hertha from Girton College to the very different context of Finsbury Technical College and, later, the Central Institution in South Kensington, this chapter will explore the culture and practices of practical science to illustrate the difficulties, both conceptual and actual, presented at turn-of-century by a woman practising in a male scientific domain.

In the last chapter, a discussion of Grace's experiences at Göttingen served to illustrate how pure mathematics' flight to abstraction connected to its privileging of individual minds and ideas of elite intellect, and how both helped to create a gender bias that posed questions for women seeking to contribute on the same level as men. In contrast, in this chapter it will be suggested that in order to present themselves as modern and progressive, scientists at Finsbury and South Kensington emphasised action in the world and cultivated an active, masculine concept of service to legitimate their concerns. On one level, Hertha was welcomed into the technical community in accordance with its self promotion as a new and meritocratic movement in contrast to the elitism and old-fashioned assumptions of Cambridge science and much of the Royal Society. On another, negotiating as they were for status and finance, these new technical, education and engineering initiatives could not afford any hint of femininity to cloud their image as producers of 'heroic' men who embodied a new kind of scientific citizenship, and whose skills were to be used for the good of society, country and Empire. As Hertha strived to create an identity as a professional scientist 'from the scientific, not the sex, point of view',[6] so sympathisers, observers and detractors sought to foreground her femininity.

Descriptions of Hertha's public lectures and experiments echo the ETU banner by using language which elides a woman in command of nature with woman *as* nature. In June 1899, the *Daily News* reported how she invoked 'astonishment' by bringing unpredictable forces to heel at a Royal Society Conversazione where she was 'in charge of the most dangerous-looking of all the exhibits – a fierce arc light enclosed in glass. Mrs Ayrton was not a bit afraid of it'.[7] With similar admiration for a wonder of scientific femininity, a distinguished electrical engineer used lyrical, rather than scientific, language to describe the 'beautiful experiments' on wave formation by which Hertha made herself 'mistress of sand ripples'.[8] In these descriptions Hertha is presented as an extraordinary individual far removed from the prosaic abilities and concerns of her more ordinary sisters. Unable to mould her into expectations of what a (male) engineer and scientist should be, and unwilling to accept her as no different from her male peers, commentators turned her instead into an iconic 'angel of science'. This kind of characterisation worked to Hertha's advantage while she was young and at the start of her career but, as will be discussed in Chapter 7, it easily became inverted as she became an older widow with political as well as scientific ambitions. At Göttingen, ideas of abstract, male genius worked to keep gender hierarchies in place in the context of pure mathematics; at Finsbury College and the Central, it was the fusion of masculinity with a new kind of scientific citizenship that combined to keep Hertha and other female practitioners at the periphery of the engineering profession.

Women, technology and technical training

An important factor preventing women from participating fully in male-dominated technical areas has been their lesser access to training. Even when opportunities are open nominally to both sexes, the culture of the training and education environment, together with gender prescriptions of society at large, can combine to create a strong disincentive to technical women. When Hertha embarked on a course of evening classes at the new City and Guilds Technical Institute at Finsbury in 1884 she was one of just three women taking electrical and applied physics courses alongside 118 men.[9] Most women, although still small in number, preferred applied art and combined evening classes with occupations such as teacher, clerk or illustrator. Hertha's decision to study at Finsbury was a reflection of her discontent with the tutoring roles that she had undertaken since leaving Girton, and may have represented for her an opportunity to add expertise to inventive skills which had already resulted in patents for a 'mathematical, dividing and measuring instrument'.[10] This small device was manufactured by W.F. Stanley of Holborn and marketed as being useful to architects and engineers for such tasks as specifying 'treads and risers of stairs, joists, roof-timbers, girders, brick spaces ...'.[11] Although the fact that a woman had solved 'a problem that has often taxed the ingenuity of technical men' caught the eye of the press,[12] the design and manufacture of small mechanical instruments was not a novel activity for a woman. Women were active in a thriving scientific instrument trade, concentrated on the fringes of London, which offered bespoke and standard products and sought custom from institutions, laboratories, professional practices and individuals. Discounting the 'invisible' women who worked with male relatives in small businesses, women were registered in their own right as makers of drawing and mathematics instruments and producers of navigation and optical equipment.[13] Hertha may well have harboured ambitions to earn an independent living in this way, following the example of her late watch-maker father, a suspicion supported by her choosing to study at a college that was newly established to provide practical training that could be transferred to the workplace.

Finsbury Technical College had been conceived as part of a broader initiative to address the scarcity of structured technical training amidst fears that other countries, notably Germany and America, would outstrip England in commercial competitiveness. In 1879 a committee set up by the City and Guilds of London Institute had established evening classes in chemistry and physics in premises rented from a school. These classes were soon oversubscribed and facilities judged inadequate, so a new college was built at Finsbury in 1883 to provide a permanent home. Funded by the London livery companies, the College's emphasis was placed firmly on training for industry. As one of the first professors, John Perry, explained in an address to the Old Students'

Association, Finsbury's training was 'peculiarly practical'. He stressed that although managers could be prejudiced against college-trained (as opposed to apprentice-trained) students, 'they would show them that Finsbury is a college of a different class, which can turn out men who are really worth their salt, who will have some sympathy with workmen, and who will not be afraid of dirty hands and hard work'.[14] Efforts were made to build connections with local industry so as to secure students and credibility; apprentices, journeymen and foremen who were employed during the day took evening classes, while individuals aged fourteen years and over attended sessions in the daytime. Electrical engineering was a speciality at Finsbury and other subjects included chemical and mechanical engineering, cabinet making, building trades and applied art; advertised on the timetable for the year that Hertha attended was an evening class on steam engines. After two years, or three in the case of evening students, a 'Technological Certificate' was awarded to successful candidates.

The austere, factory-like architecture of Finsbury College reflected its role as a 'hands-on' servant of industry. Unplastered walls served to highlight its affinity with the workshop and there were no reception or committee rooms, library, refreshment or staff common room. In keeping with this masculine, industrial arrangement, the original plans had omitted any female toilets – an oversight that had to be remedied hastily when a need was recognised.[15] This environment was far removed from the gentility of Girton College and the comfortable drawing rooms that Hertha had frequented as a private tutor. There was no room for overt femininity (or comfort) in a college designed to counter scepticism of engineers trained away from the workplace, or in one which aimed to prove that 'Finsbury men' were just as tough and willing to get their hands dirty as apprentice-trained technicians.[16] That women might want to engage in such a manly, utilitarian activity was viewed with amusement and indulgence by *The Electrician* magazine which remarked that at Finsbury 'Women may study electrical science without risk of alarming anybody or of doing any harm to themselves'.[17] Hertha did so three times a week for a year, funding herself with the financial support of her Girton benefactor Barbara Bodichon. Student records show that Hertha's choice of 'electrotechnics' was by far the most popular course attracting more than twice the number of any other. As the omission of a 'ladies' in the design of Finsbury College suggests, women were not seen as potential students in any number – after all, women were not admitted to most trades and Finsbury trained individuals for industry.

Hertha concluded her training at Finsbury in 1885; her name (Sarah Marks) appears among 115 men in the table of results for that session at an undistinguished place near the bottom. For her laboratory assessment Hertha had gained just seven marks, significantly less than the top mark of seventy-eight.[18] Although experimentation was an important component of the Finsbury training and 'for every hour that a student spends at lecture, he spends several in the laboratory',[19] it seems that Hertha was one of a number who did not attend the full assessment or failed to deliver proper 'laboratory notes'. This poor performance is indicative of the problems faced by a rare woman among the many men in the crowded electro-technics laboratory at Finsbury. Here, many of the students would be partially familiar with the workshop apparatus of telegraphy or electric lighting from exposure to them in their 'day jobs', an advantage that Hertha did not possess. Furthermore, students worked in teams to pursue their experimental work and it would have been difficult for a relatively inexperienced woman – at thirty years of age older than many of her fellow male students – to become 'one of a team'. (Hertha's age is another indication of the particular way that women accessed technical education, coming to it via other routes rather than following a direct path via school, apprenticeship or the workplace as did most of the men.)

Electro-technology courses were over subscribed at Finsbury and Hertha may have experienced tension in competing with men for apparatus, space and a glimpse of the demonstrator in crowded and, at times, disorderly classes. Although matters were reported to have improved to an extent by 1884, the previous academic year a student had complained to *The Electrician* that Finsbury was 'little better than in a state of chaos' and that the chief demonstrator was 'principally occupied in taking tickets and names at the door'.[20] At this time at Girton, women were still required to be accompanied by chaperones to outside lectures and coaching sessions with male tutors (this requirement would persist, nominally at least, until the next century) and it is difficult to see how Hertha could have made best use of Finsbury's training, jostling with men for space and resources, and have adhered to middle-class codes of genteel, female propriety. It is not surprising that even Sharp, in her proselytising *Memoir* designed to illustrate a woman's capability in a 'man's world', pays scant attention to Hertha's experiences at Finsbury, concentrating instead on her courtship with her professor there, William Edward Ayrton (1847–1908) and the latter's feminist credentials.

A marriage of scientific equals?

One way for a student to cope with the masculine classroom dynamics at Finsbury would be to have the special, out of hours, attention of the teacher. After a brief courtship, Hertha and Ayrton were married in May 1885. By this time, Ayrton had already achieved a reputation in telegraphy, electrical engineering and technical education, and in 1881 had been elected a fellow of the Royal Society. After studying mathematics at University College London, Ayrton had worked in the physical laboratory of Sir William Thomson (Lord Kelvin) at Glasgow, before taking posts with the Indian Telegraph Company and Great Western Railway, all of which gave him strong industrial credentials. Ayrton had returned from heading up the Department of Natural Philosophy and Telegraphy at the Imperial Engineering College in Tokyo to take up his professorship at the City and Guilds Institute (1879) and Finsbury Technical College (1881).

It has become almost orthodoxy in the history of women in the sciences to point out how women needed 'male mentors' to facilitate access, create opportunities and act as intermediaries with the male scientific establishment. Taking this a step further, there is within this trope a developing body of scholarship illustrating how marriage was a significant route into science for many women.[21] However, as will be illustrated here with the relationship of Hertha and Ayrton (and with that of Grace and Young in the succeeding chapter) the 'mentor' model can sometimes be a limiting tool with which to analyse women in science. It may become a stereotype which obscures as much as it reveals, hiding the complexity of women's negotiations within science and implying passivity and a lack of agency. Although there is no doubt that Ayrton was an advantage to Hertha – as she surmised on her engagement 'he is going to let me go on with my electrical work, and of course he can help me in it in every way'[22] – the advantage also flowed in the opposite direction. Hertha was an asset in Ayrton's bid to establish his modern, emancipated credentials as a progressive man of science, and their scientific collaborations reveal a carefully-planned public relations strategy designed to further both their reputations. In addition, most of the little scholarship on Hertha focuses on her researches on the electric arc, yet this was not an all-defining work and Hertha's experimental career continued for a full fifteen years after her husband's death.

There is little doubt that Hertha was a suitable consort for Ayrton. Together they were a feature of London scientific society; attending

Friday evening lectures at the Royal Institution, hosting a luncheon party for Heinrich Hertz (who had travelled to England to receive a Royal Society medal), presiding over Finsbury College's annual 'old students' dinner', or attempting to discover the science behind the 'thought-reading' couple who were 'bewitching' London at the Alhambra Theatre.[23] Ayrton's first wife had died two years previously; she too had been an independent-minded individual and was one of the first English women to gain medical qualifications.[24] When Ayrton died in 1908, appreciations were unusual in that they gave almost as much attention to his views on sexual equality as they did to his scientific achievements. A funeral notice in the *Morning Post* recalled how 'Ayrton felt that science was of no sex, and that chivalry could not consist of opening the drawing room door for women and closing the doors of scientific society'. The obituary continued with a call to Ayrton's peers: 'If his fellow-workers in science wished to honour his memory they could not better do so than by following his inspiration in this regard'.[25] A few years earlier Hertha had been refused a fellowship of the Royal Society on the grounds of her sex and married status; this comment may well have been directed to the fellows and members of council who had refused her entry. In similar vein, a letter to *The Times* noted Ayrton's 'affinity for intellectual womanhood' and made the point that, unlike some men of science, he did not absorb his wife's life and work into his own. 'On the contrary he exerted himself to have her career recognised as separate and individual. This was his real contribution to suffrage.'[26]

Ayrton's sympathy with the women's cause was shared by other of his engineering colleagues. John Perry twice nominated Hertha for a Royal Society medal (and wrote her statements of endorsement) and Thomas Mather collaborated with Hertha and was one of her most loyal supporters.[27] On Hertha's death, Maurice Solomon, editor of the *Central Gazette* (the College newspaper) recalled that Hertha had been an 'inspiration' to the students who had worked with her. Despite this, he continued, Hertha's sex had been 'a continual hindrance to her work' and she had had to withstand 'a constant struggle against prejudice and old-standing prohibitions ... which even in success had led too many to judge her achievements by the sex of the doer'.[28] This public support for equality of opportunity and the suffrage cause can be interpreted as part of the negotiations of this new, practical science and education initiative to secure credibility for its methods and aims.

During Ayrton's lifetime, the suffrage campaign had remained fairly 'ladylike'; although the militant Women's Social and Political Union

(WSPU) had been established in October 1903 (with Hertha as an early member) it was not until around 1910 that its militancy escalated. As a result, the suffrage movement split into militant and non-militant wings and, for professionals in the public eye, arguing the militants' cause became considerably more controversial. Despite this, especially in the 1900s, supporting equal rights for women was in keeping with the presentation of these professional men of science as the prototype of the modern, progressive citizen – a new vision that stood in opposition to the 'prejudice' and 'old-standing prohibitions' of the older tradition of gentlemen of science. Tension between these two groups was played out on various levels including the nature of scientific education, the dangers of commercialism, and the place of women in science. In Chapter 7 it will be demonstrated how Hertha's nomination to the Royal Society was part of an ongoing battle between opposing factions within that learned organisation; here the focus will remain on the Central Institution where Hertha carried out her best-known investigations into the electric arc.

Central men, active masculinity and scientific citizenship

The Central Institution at South Kensington opened in 1884 with financial sponsorship from the City and Guilds of London. It was renamed the Central Technical College in 1893 and in 1907 became a constituent of the newly-formed Imperial College of the University of London. The Central catered for 'graduates' from Finsbury and other schools/ universities (Ayrton was particularly gratified to have a Cambridge wrangler in his class of 1896), returnees from industry and a growing number of overseas students, especially from Japan, India and Egypt. These individuals came to the Central to gain advanced technical and professional training and the College is generally seen as being a prototype for the later polytechnics.

One historian has written that by 1881 English technical education was reaching its 'heroic stage'.[29] The gendered adjective 'heroic' is telling. Pure mathematicians, and other men engaged in inactive intellectual pursuits, at times felt a need to reinforce and display their masculinity by undertaking rigorous 'manly' physical activity. Bookwork could be seen as feminine, even effeminate, and around 1900 there was a vogue among male intellectual workers for demanding activities such as mountain climbing and marathon walking.[30] By the 1890s, a tension has been identified within elite masculinity which opposed the qualities of physicality and intellectualism, privileging the former as the defining virtue of the elite male.[31] Linking to this, John Tosh has argued for a late-Victorian model of manliness which, for its realisation, required a forceful exertion of will over wives.[32] In the context of science, there are several examples of collaborative partnerships in which joint work or the work of the female spouse has been presented as solely the man's responsibility. The succeeding chapter will demonstrate this dynamic with regard to Grace Chisholm and William Henry Young; Barbara Becker has illustrated a similar arrangement between William and Margaret Huggins.[33] That Ayrton and other Central men's attitudes were altogether more 'advanced' rested in part on the knowledge that they were secure that their own qualities, and their profession, were gendered unquestionably masculine.

William Ayrton had proved his expertise and stamina by helping to transform the Indian telegraphic network and setting up his own telegraphic laboratory in Japan. Often referred to as 'the nerves of Empire', Britain's global telegraphic network had a moral subtext and its engineers were celebrated for their service to imperialism and their manly courage. One example of this is a series of cigarette cards on 'wireless telegraphy' which were issued by Lambert and Butler in 1909 and emphasised the severity of the conditions that telegraphic engineers faced in their work. At both Finsbury College and the Central the emphasis was placed firmly on active, practical learning – not feminised, sedentary bookwork. As early as 1886, assumptions that implied just such a gender division had been articulated:

> in an age when the wonderful growth of physical science and the absorbing demand of material interests are more and more engrossing the thoughts and minds of educated men, it is to devolve on women in some way to supply the loss, and to aid in preserving and transmitting to the civilisation of the future an element of refining culture which it can so ill spare.[34]

Masculinity was also assured in the securing and use of new, often large scale, equipment and instrumentation. At the Central, Ayrton could display his potent masculinity, not least in the new electrical laboratories that he had specified and fought to have financed. Professor Ayrton had to be 'congratulated', said an editorial in the College journal, before going on to describe the new facility which promised to maintain the Central's, and England's, precedence in the fast-paced world of electrical engineering. The laboratory signalled status, modernity and scale. It was eighty-five feet long, fifty-three feet wide and contained, amongst other equipment, a power generation and distribution centre, a five kilowatt

Ferranti transformer, and a travelling overhead crane to assist in experimentation.[35]

Cultural representations of scientific manliness became more common around turn-of-century. J.A. Kesiner has argued that the Sherlock Holmes stories, which first appeared around this time, both reflected and constructed a script of active masculinity which aligned manliness with rationality, scientific procedure, observation, factuality and self-help.[36] Conan Doyle admired the work of Edgar Allen Poe, who is generally credited with originating the genre of the detective novel with his 1841 story *The Murders in the Rue Morgue*. Conan Doyle praised Poe (and later French novelist Jules Verne) for producing 'credible effects' within 'incredible story lines' through the use of a knowledge of nature.[37] Although it is certainly the case that Poe's detective, Inspector Dupin, anticipates the character of Sherlock Holmes in the use of analytical reasoning,[38] the later Holmes stories place scientific procedure as essential and constitutive of the Holmes's character. In this way, Conan Doyle created a combination of the 'detective genre' with that of the 'scientific expedition'.[39] H.G. Wells, who had been taught biology and zoology by Thomas Huxley at the Normal School of Science,[40] also presents the scientific man as an ideal of masculinity in his novel, *Ann Veronica*, which is based around the natural science laboratory at Imperial College (of which Huxley's College became a constituent). Capes, the rational, plain and trustworthy scientist is contrasted with his foppish, effeminate rival who feels uncomfortable in the laboratory.[41] Although the heroic myth of the Victorian engineer had become a part of the legitimising myths of industrial society earlier in the century (reflected, for example, in the biographies produced by Samuel Smiles)[42] this later characterisation was different in the emphasis it placed on scientific method as applied to all aspects of society and as a paradigm of manly citizenship. A leader in *The Times* welcomed the proposed establishment of Imperial College because it promised 'science systematically applied to the arts and industries of daily life'. The writer envisaged that the new educational establishment would become

> a model and a centre of light leading the whole Empire ... We talk a great deal about science, and we all pay it a certain amount of homage with our lips; but we still withhold from it in too large a measure the homage of our heads and understandings. We have not yet learnt to see in it the influence, the spirit, and the force which really lie at the root of all effective action in the modern world.[43]

Social reformer Beatrice Webb, who was active in the scientific circles of Thomas Huxley and others, recalled that these men of science held an 'almost fanatical faith ... that it was by science and science alone that all human misery would be ultimately swept away'.[44]

For *Nature*, commenting on a report by the British Association for the Advancement of Science, the urgent issue was industrial competition between nations and this took on a military flavour (the soldier, like the engineer, was indisputably male): 'we are in the midst of a struggle in which science and brains take the place of swords and sinews; the school, the university, the laboratory and the workshop are the battlefields ...'.[45] It was this kind of thinking which informed one level of support for eugenics, a movement which only gained momentum in the years preceding World War One. Hertha was present, alongside scientific notables Norman Lockyer, William Ramsay, Arthur Schuster and Archibald Geikie, at the inaugural International Eugenics Congress held in the Great Hall of the University of London for five days in 1912. At the congress dinner, they heard principal speaker Arthur Balfour argue

> '... that a feeble-minded man, even though he survive, is not so good as the good professional man ... In any case we are scientific or we are nothing ... I am one of those who base their belief in the future progress of mankind, in most departments, upon the application of scientific method to practical life'.[46]

For Grace and her husband, eugenics was a way to ensure the continuance of an intellectual elite; for Hertha and her peers in practical science, it was about using science as an instrument to bring about a more rational, efficient society. Norman Lockyer was seeking to realise similar ambitions when he established the British Science Guild in 1905, an association of professional scientists, including Ayrton and others, concerned to promote the role of science in society.

The Central Institution

The professors at the Central were keen to present themselves as part of this modern, meritocratic scientific movement not least because it enabled them to build credibility at a time when their material practices and their integrity were not yet fully accepted. The College was not immediately successful in attracting students and it was not until the 1890s that the situation improved. Furthermore, the practical methods of the College, especially as applied to mathematics, were initially a subject of controversy which even reached the pages of *The Times*. At Finsbury, John Perry had introduced the rounding off of decimals to approximations in recognition of the needs of engineers and pioneered the use of squared paper to facilitate easier production of results.[47] The Central's adoption of a similarly pragmatic approach was not uncontentious. In the opinion of one letter writer to *The Times*, 'practical mathematics' was an innovation suggesting 'an appalling depth of ignorance', to which a colleague of Ayrton's at the Central, chemist Henry Armstrong, responded that the correspondent simply could not 'understand the modern tendency to be practical'.[48] The merits of practical versus pure mathematics were a topic on the letters page again in 1911. Henry Spooner suggested that the differing approaches of the universities and the engineering colleges on this matter accounted for the small number of university-matriculated students entering engineering and that 'most engineering colleges have for years wisely realised that mathematics, instead of being an abstract discipline, should become the means of applying truths to the everyday affairs of the world in which we live'.[49] The opposing view was put by Grace's husband, William Young, who wrote that 'to substitute the mathematics of ... the engineer for that of the expert mathematician would resemble the behaviour of a man who should insist on having a bad watch because of the variations in the equation of time'.[50]

There was a moral dimension to this dispute, centring on how the world should be properly represented and measured, and who was best qualified to do so. Was the real bringer of truth the engineer, or the exponent of a purer, more precise mathematics? Tensions of class can also be discerned, as the 'practical' men argued against the long-standing hierarchy that, in England at least, associated manual labour with the lower classes and intellectual labour with the elite. The case of the engineer to be the embodiment of morality and truth could be further undermined by those adhering to an older notion of the scientist as a disinterested investigator seeking knowledge for its own sake, not for any reward that it might bring. Ayrton was dismissed by one of his professorial colleagues as 'partial', 'impelled into science through contact with Sir William Thomson', and 'a worker chiefly at its technical and commercial fringe rather than its depths'.[51] Both Ayrton and Perry worked as practising engineers, accepted consultancy work and patented many instruments and Ayrton was aware that his interest in business was, for some, at the least a distraction if not a problem for his work.[52]

To argue that the presentation of a modern, meritocratic professionalism facilitated the professors at the Central in their support (vocally at least) for women's equality seems to be to deny the explanatory model of the increasing professionalisation of science as a barrier that kept women out. This process is often linked to Darwinian/evolutionary influences amongst the new professionals which led them to discount women's capacity for scientific pursuits. This conviction was certainly held by sections of the scientific community, including professor of chemistry Henry Armstrong. However many other 'scientific naturalists' did not make this connection and staunch Darwinists such as Ayrton, Hertha, and editor of *Nature* astronomer Norman Lockyer, took an opposing view. Despite being a strong explanatory tool, the professionalisation model does not account for the vocal support for women exhibited by many sections of the science and engineering community. Lockyer campaigned in his journal in support of the admission of women to learned societies and for the equal treatment of intellectual work, irrespective of the sex of the producer. When Hertha was proposed for a Royal Society fellowship, nine prominent male fellows signed and supported her nomination. When Armstrong used his report for the 1904 Mosely Education Commission to argue against women's intellectual equality and for the 'ruinous effect' of female science teachers, an editorial in *The Central* contradicted him decisively: 'Professor Armstrong's argument from evolutionary principles seems to us absolutely fallacious – he would probably say that this is due to our inability to appreciate them'.[53]

It seems clear that the processes of professionalisation need to be unpackaged in order to reveal the complexity hiding behind the term. The new engineering and academic professors were keen to distinguish themselves from the elitist ideology and exclusive practices characteristic of much of Cambridge-dominated science and of traditional sections of the Royal Society.[54] To support women in the interests of 'fairness' and 'justice' was logical and supportive of their cause, and it was from similar principles that many of these scientific 'modernisers' backed votes for women. This is also an important factor in the public relations strategy of Hertha and William Ayrton. Both were keen to deny in public any hint of collaboration or collusion in their work, only too aware of the gendered interpretations given to collaborators of differing sex. As Hertha wrote of Margaret Huggins 'she has done some splendid work on astronomy herself, with her husband, and has not had a bit of recognition for it just because no one will believe that if a man and a woman do a bit of work together the woman really does anything'.[55]

The politics of collaboration

Yet the Ayrtons' public disavowal of collaboration was not entirely true in reality (and collaboration is, after all, a flexible term that can encompass a myriad of types and extents of shared involvement). For example, when ill health struck Ayrton in 1903, just as he had been commissioned to report on searchlights for the Admiralty, Hertha increasingly took on the major part of the work, experimenting at the Central with her husband's students and staff.[56] The research resulted in four reports, three of which were published under Ayrton's name alone. The work also resulted in an article, attributed to Hertha, in the *Times Engineering Supplement*. In addition, Sharp alludes to Ayrton contributing to Hertha's papers[57] and a colleague at the Central remembers that 'being both enthusiastic and having cognate interests, they constantly worried each other about the work they were doing'.[58] Hertha's important investigations into the electric arc were a continuation of her husband's work; her 1903 book which resulted from these researches illustrates clearly the assistance received from staff at the Central, in particular Thomas Mather who made fundamental contributions to the design of the experiments.[59] The attributing of credit and 'ownership' of research is a complex phenomenon and recent scholars have debunked the idea of a single, heroic, lone experimenter. Who 'takes the credit' for a discovery or piece of research can be a political, as well as a scientific, issue; that women have been classified as 'assistants' has been well documented by historians of women and science.

Hertha's researches on the electric arc were carried out in her husband's laboratories at the Central Institution. Arc lights were used for searchlights, streetlights and other public lighting, as the light that they produced was too harsh for indoor home use. To strike the arc a high voltage was established between two carbon rods a short distance apart; however the result was unstable and there were additional problems of noise, sputtering and hissing. By experiment and observation, Hertha discovered that one of the conditions of the steady burning of the arc was the shape assumed by the ends of the carbons; by shaping these to a certain dimension the arc's performance was improved considerably. Her major theoretical insight was to explain the hissing and instability of the arc as a result of air rushing to the tip of the positive electrode, oxidising the carbon and forming a crater. This led her to establish a relationship between the voltage drop across the arc, its length and the current that became known as the 'Ayrton Equation'.[60] Despite this lengthy and well-received investigation, Hertha had no official role or status at the Central Institution and her presence was dependent on her being 'wife of the professor'. When her husband died in 1908, she had no further dealings with the College and, instead, enlarged her home laboratory to better cater for her needs. However, the loss of a viable and credible experimental space was to have serious implications for the reception of Hertha's work, as will be demonstrated in Chapter 5.

Femininity confined to the margins

Despite the vocal sympathy for women's equality in the sciences displayed by important sections of the science and engineering community, Hertha encountered obstacles in attempting to negotiate an identity as a professional engineer like, and alongside, her male peers. Although the professors may have been supportive in principle (and indeed the College's 1878 foundation documents show that the City and Guilds initiators planned an institution catering for both sexes),[61] the later aims and material practices of the College quickly militated against this agenda.

The priority of the Central was to train professional men and become the equal in prestige of the more established universities. Women, especially middle-class women, were still associated with amateur status and the acquisition of learning for learning's sake; as a result they became marginal to the College's preferred image and, arguably, threatening to College aims. Indeed, women were rarely recruited to the Central and no names recognisable as women's can be found on existing engineering student registers to 1899.[62] Despite Hertha's periodic researches at the Central from around 1895 to 1904, which included her well-known investigations into the electric arc, her presence is rarely noted in any extant College archive. Hertha is mentioned briefly as a collaborator in an obituary of her husband in the college journal; after her death this also carried a review of Sharp's *Memoir* published in 1926. This relative 'invisibility' is significant when it is remembered that Hertha's work there, aided by Central students, led to the award of a prestigious Royal Society Medal (awarded in part for a later investigation into sand ripples). Despite Hertha's informal status at the Central, this was surely something to publicise for a college striving to win credibility as a professional institution offering a university-equivalent education? Indeed, promoting the original work of its students was an important propaganda tool for the Central in its bid to win credibility and prestige.[63]

This reticence over Hertha's links to the College, and the absence of any significant number of women students, is to be explained in the context of the Central's primary purpose, as outlined in its 1884 Scheme for Organisation. This 'mission statement' stressed that the College's aim was 'to point out the application of different brands of science to various manufacturing industries' noting that 'in this respect the teaching will differ from that given in the Universities and in other institutions in which science is taught rather *for its own sake* than with the view to its industrial application' (my italics).[64] Students of the Central were being trained to take jobs at middle to senior manager level in electrical, construction and other industries, and could be sent to the furthest reaches of Empire. Just as the heroes of popular Edwardian novels were presented as achieving manhood by leaving women and home to travel overseas and test themselves in the comradeship of men,[65] so 'graduates' of the Central were prepared for taking on strenuous and demanding tasks, often in foreign lands. The College had its own Employment Bureau and, once armed with their Diploma, 'Central men' took posts with the Royal Survey, electric supply and railway companies, lift manufacturers, dam construction companies and similar organisations.

Within this self-consciously manly environment, even the merest hint of femininity required explanation. When *The Central* described the College's research on a vacuum cleaner, the author felt it necessary to excuse this feminine-related subject 'occupying valuable space' by emphasising that 'the profession of an engineer is the art of directing the sources of power in Nature for the use and convenience of men'.[66] The Central did not even train individuals to be teachers (an acceptable role for women) as that was the task of its partner college in South Kensington, the Royal College of Science. Women were more visible here (especially in seasonal summer courses) but, as noted in a survey of opportunities for women published in 1897, 'a fair number of women have now received an excellent training in natural science, holding the BSc of Cambridge or London (but) hardly any employment is open to such women save that of science mistress'.[67] A woman studying science *for its own sake* was one thing, a woman being trained to apply science to industry was quite another.

Yet the rarity of women in either of these spheres could easily fuel an attack on a woman's femininity and sexual desirability. By the later 1900s women were more visible but, as a poster/cartoon illustrating different kinds of female university flappers indicates, science and medical students were still viewed as odd (Figure 3.2). As opposed to the young and nubile 'Slade' and 'Arts' flappers, the science and medical women are portrayed as aged and unattractive; for science, the word 'flapper' is in quotation marks as if to imply that it was not really possible to be a science student *and* a young, modern, desirable woman. It will be discussed in Chapter 7 how older, menopausal women became figures of ridicule – even of hate – among sections of the medical and general community; presenting the 'science flapper' as aged may have been another way of

Figure 3.2 Cartoon from the University College London *Union Magazine* (1915), by kind permission of College Collection, UCL Library Services, Special Collections

highlighting her ridiculousness. The representation of the 'medical flapper' is also unflattering, even menacing; she is portrayed as a dark, gaunt figure with the large nose and abundant hair (caught up in a scarf) suggestive of stereotypical Jewishness. The scalpel held in her hand suggests she is about to perform a (diabolical?) dissection. Fear of the Jewish 'other' has been discussed in Chapter 1 with reference to Hertha's experiences at Girton. It has also been noted that the emergence of an increasingly successful, educated and financially-comfortable Jewish community constituted a challenge and source of unease to sections of Anglican England. The figure of the Jew was 'constantly invoked at this moment when Christian authority was under heightened scrutiny in an era of religious scepticism due to Darwin'.[68] The image of the 'medical flapper' seems to combine this fear of the Jewish 'other' with anxiety over women stepping beyond their prescribed roles and competing in 'men's' scientific arenas. This kind of representation would have conveyed worrying subliminal messages to middle-class Jewish women such as Hertha and may be another clue as to why she felt it necessary to marginalise her Judaism on her marriage.

Engineer or lady of science?

Hertha has been remembered as a 'persistent experimenter'.[69] She enjoyed being in the laboratory carrying out practical investigations and consistently applied her inventive and engineering skills to the development of devices and equipment. As she confided to a relative of Marie Curie, when working in her laboratory she felt that for her 'la science est toujours la grande calme, un réfuge contre tout les maux'.[70] Her first design for a sphygmometer (device for measuring the pulse) was produced while she was still a student, in later life she designed various devices including an anti-gas fan for use in the trenches during World War One; her final years were spent in trying to adapt this for application to ventilation in ships. From her letters, her inventiveness and the many patents that she took out, it is clear that she would have liked to lead a career much like her husband's, but the identity of a modern, professional engineer was indisputably male, despite women's long (and until recently) largely invisible history as inventors and producers of technology.[71] The engineers at the Central contrasted themselves with the older figure of the disinterested research scientist, and the newer pure mathematician lost in 'useless' abstraction, by dint of their active, manly citizenship. Action in the world was central, based on empirical knowledge, measurement, trust and a rejection of any science based on metaphysical interpretations.

Londa Schiebinger, Evelyn Fox Keller and others[72] have demonstrated how, in the seventeenth century, 'femininity' became the antithesis of the new, virile 'masculine' science of the emerging Baconian Royal Society. 'Feminine' became descriptive of a contemplative, deductive, passive style of investigation that was to be expunged from the new values and epistemological stance of modern, active, experimental science. A parallel process can be discerned at the end of the nineteenth century as the new, thrusting professional lobbied for credibility with his purer, elitist rival. But more than this, the modern engineer applied his science for the betterment of the world and sought legitimacy for his science (which to the old school was tainted by a compromising commercialism) with ideals of service. Their science was not only to be put at the service of men, but of country too. For example, at turn-of-century Central students created 'The Electrical Engineers' voluntary corps, sanctioned by the War Office, which concerned itself with electricity as an agent of warfare.[73]

Notions of service had, since the mid-Victorian period, strong feminine connotations; women's ideal role was to remain in the private sphere supporting their families and communities by performing the role of 'helpmeet' and undertaking philanthropic acts. The engineers sought to put a masculine (and at times military) face on an ideal of service that was directed out into the world; a woman in their midst may have been compromising to this image. The ETU banner, bearing the words 'light and liberty', showing men in the their work attire at the service of an abstract and female electricity/nature, conveys neatly the gendered roles within this new profession of engineering. It also suggests why it was so hard for practitioners and observers to accept Hertha as a scientific professional no different from her male peers. With few examples or narratives of feminine technical or engineering professionalism with which to align a female scientist, Hertha was instead classified with the other 'female' element of the ETU banner – the extraordinary angel or goddess of natural forces, an elision of woman scientist seeking to control nature *with* nature itself.

The personification of nature in female form is a long-standing phenomenon. In her definitive text, *Monuments and Maidens*,[74] Marina Warner finds a prosaic reason for this kind of female allegory in the common relation of abstract nouns, especially of virtue, to the feminine gender in Indo-European languages. However, her more profound analysis reveals that recognition of the symbolic order, inhabited by ideal, allegorical figures, as opposed to the actual order, is often dependent on the unlikelihood of women practising the concepts that they represent. And (as the example of the ETU banner with its depiction of real men and abstract 'woman' demonstrates) the female form tends to be perceived as generic and universal, the male as individual, even when expressing a generalised idea.[75] Londa Schiebinger has applied this analysis to representations of science and has demonstrated just how common female personifications of science, and branches of science, were in the seventeenth and eighteenth centuries.[76] These images were often equipped with the latest scientific accoutrements, such as a barometer, telescope or vacuum jar, and appeared typically on the frontispiece of scientific texts. In this way, science was pictured as feminine not least because the scientists, 'the framers of this scheme' were male and 'the femme Scienta plays opposite the male scientist'. Such images also buy into the discourse of 'the muse' as 'the scientist imagines that a feminine science leads him to the secrets of nature or the rational soul'.[77] But when the scientist is female, this arrangement has serious implications.

Despite her attempts to present herself as a professional scientist, Hertha's femininity was foregrounded constantly and she was often typecast as a supernatural 'heroine' of science. A one-time president of the IEE used lyrical in preference to scientific language to describe her 'beautiful experiments' on wave formation, recalling how Hertha made herself 'the mistress of sand ripples' and describing how the movement of sand 'at her will formed itself into beautifully uniform patterns'. Adding to the image of Hertha as some extraordinary example of womanhood, A.P. Trotter continues to tell how he read one of her papers to his students and showed a lantern-slide portrait of the author in an attempt 'to describe her charming personality'.[78] The difficulty of accommodating a female scientist within any other, more prosaic narrative, is illustrated by a popular text, *Woman in Science,* which was published in 1913 and went through a number of editions. In his preface, H.J. Mozans describes how his mind turned to women during leisurely wanderings through the 'famed and picturesque land of the Hellenes'. Here, inspired by representations of Aspasia 'the virgin goddess of wisdom and art and science', he was moved to investigate the intellectual achievements of women. In Mozans' flowery prose, descriptions of real, nineteenth-century scientists are often couched in language more usually descriptive of mythical female deities. Entomologist Eleanor Ormerod is reported as being described by the university official who conferred an honorary doctorate upon her as 'entitled to be hailed as the protectoress of agriculture and the fruits of the earth ... a beneficent Demeter of the nineteenth century'.[79] In similar vein, Marie Curie's experimental abilities are likened to magic as 'before her deft hands and fertile brain difficulties vanished as if under the magic wand of Prospero'.[80]

The difficulties of a woman taking on the new – but ordinary – role of a professional scientist at turn-of-century can be gleaned from Mozans' concluding remarks. Here he generalises that men will 'continue as specialists as long as the love of fame, to consider no other motives for research, continues to be a potent influence in their investigations', while women will forego 'long and tedious processes' for the 'proper apprehension of higher and more important truths'. This was in keeping with a thread within contemporary thought which pointed to woman's greater purity and nobility and stressed her role in tempering the excesses of man.[81] For Mozans, men reached conclusions through the application of plodding, inductive methods, while women naturally used deduction and 'a kind of intuition, coupled with (their) more pronounced idealism'. As a result

> ... just at the critical moment, when men of science would rather discover a process than a law, when they are so preoccupied with the infinitely little that they lose sight of the cosmos as a whole ... when, like Plato's cave men, they have so long groped in darkness that their powers of vision are impaired, then it is that woman, 'The herald of a brighter race', comes to the rescue ... For women, as a rule, love science for its own sake ...[82]

It is interesting that Henry Armstrong's obituary in *Nature* criticised Hertha within just these terms of reference, maintaining that 'though a capable worker, she was a complete specialist and had neither the extent nor depth of knowledge, the penetrative faculty, required to give her entire grasp of her subject'.[83] Hertha's refusal to be placed on a pedestal of science, preferring to be a professional worker rather than a sanctified, exceptional 'mistress' or 'lady' of science, may have been part of the problem for her unsympathetic obituarist. After all, if access to science was conceived as possible for only special women – women exceptional within their sex – then female scientists would be rare; open the doors of science to women as everyday professionals and significant numbers of 'everyday' women may enter. Armstrong held well-known views about the incompatibility of women and science and was vocal in his opposition to women's admission to the Chemical Society. He was particularly ungenerous in his obituary of Hertha, accusing her of failing in her wifely duties: 'He (William Ayrton) should have had a humdrum wife ... who would have put him into carpet-slippers when he came home, fed him well and led him not to worry either himself or other people, especially other people; then he would have lived a longer and a happier life and done far more effective work ...'.[84]

In the Mozans passage quoted above, women seem to represent an antithesis to the modern, technological and commercial concerns of practical science and engineering. Even female scientists are charged as having, by dint of their female nature, a higher calling and the ability to temper the myopic scientific tendencies of men. This signification was in keeping with a thread in late-nineteenth-century society which reacted against modernity and material growth and characterised the industrial world as alienating.[85] Femininity was often the pivot around which such concerns were articulated, as in the idealised role prescriptions for women in the work of writers such as John Ruskin and Charles Kingsley. This thread was fully articulated in Germany, where modern technology and industry were seen as the incarnation of masculine principles which subjugated individuality and diversity to standardised industrial practice. Here women pointed to their authenticity and ability to reconnect men with nature as part of their demands for increased influence and emancipation.[86] In England too, the ideal of womanhood as something pure and noble in contrast to the material, masculine concerns of the modern world was a factor in arguments against female suffrage.[87]

In attempting to find a place for themselves in science, earlier women had sometimes colluded with supernatural representations of themselves and identified with the feminine icon. Both astronomer Maria Cunitz and natural philosopher Emilie du Châtelet placed themselves amongst the muses and not within depictions of actual (male) scientists on frontispieces of their work.[88] In the nineteenth century too, women used similar methods to negotiate their position in science. Mathematician Mary Somerville preferred to remain uncontroversial and was happy to present herself as a 'lady' of science, in keeping with contemporary ideas of gendered intellect. With due feminine modesty, Somerville wrote that as a woman she was not original herself, but only an interpreter and presenter of the original work of men. Kathryn Neeley argues persuasively that this was a useful strategy for a female mathematician of her class and age, rather than a strongly-held philosophical commitment.[89] Similarly, anthropologist Clémence Royer argued that women had a natural 'genius peculiar to themselves' and that they contributed to science in a feminine way, different from men.[90] For Hertha, such a position would have been unthinkable; she argued strongly against any biological difference between the sexes and felt that it was only social handicaps that needed to be removed to facilitate women's involvement in activities dominated by men. Throughout her life, Hertha reiterated constantly her 'plea for equality of treatment of intellectual work without regard to the sex of the worker'.[91]

Conclusion

Whether researching the electric arc under the auspices of her husband's department at the Central, or in collaboration with him in the development of searchlights, Hertha strove to retain an independent identity. Despite her efforts, the marginal status of women at the Central created difficulties for her in exploiting her well-received work on the electric arc to its full potential as a stepping stone to other experimental opportunities. In keeping with her convictions on sexual equality, Hertha was a political strategist and firm defender of her work; she did not present herself as a deferential 'gentlewoman of science'. Her correspondence shows that, from the start, she was not afraid to argue forcefully for her work, for example defending her observations on the electric arc against Sylvanus Thompson's well-meaning criticism[92] and engaging in bad-tempered correspondence with referees of her Royal Society papers (as will be illustrated in Chapter 7). Neither was she afraid to argue in public. *The Times* letters page contained a firm rebuttal by her of criticism of her anti-gas fans[93] and when William Ramsay suggested in the *Daily Mail* that women scientists do their best work when working under the guidance of men, Hertha's reply was devastatingly effective – and personal:

> Collaboration is apparently not a womanly but a human characteristic ... In testifying to the stimulus provided by collaboration, Sir William Ramsay speaks from a wide experience denied to me; for most of his own work has been done 'when collaborating with a male colleague' and one can imagine the fervour with which he worked 'with Lord Rayleigh and for Lord Rayleigh' at the discovery of argon.[94]

This assertiveness is conveyed in Armstrong's obituary when he contrasts Hertha's possession of 'the vigour of Wotan's masterful daughter Brunhilde' with the (preferable) charms of Ayrton's first wife who was a feminine and 'ethereal being'.[95]

Hertha was not, however, entirely successful in adopting the identity of 'engineer' in the face of the conceptual problems facing technical women outlined above. On her death, *The Times* headed its obituary 'A distinguished *woman scientist*'[96] (my italics); this was a common representation of Hertha and one which she would usually accept. Although Hertha was an experimenter and designer of technology, engaged in engineering and invention to the end of her life (she took out patents from 1883 to 1920), the title 'scientist' could also reflect her interests. As the press preferred to report her activities under the banner of 'lady' or 'woman' scientist, Hertha may have acquiesced to this title in response to the evident tension between the term 'engineer' and 'woman'. (Indeed, this was a tension that she may have sensed herself.) This, in turn, suggests the difficulties that she faced when trying to participate in engineering at the level of professional practice. Mary Somerville perfected the role of the gracious lady of science, seldom demanding rights for herself on a par with men and happy to have her bust placed in the Great Hall of the Royal Society, even if she herself was not welcome. For Hertha, this was not enough; she wanted to be admitted in person, just as she wanted to be accepted as a working, scientific professional alongside her peers. However for a woman at the turn of the nineteenth century, there may have been room on the pedestal, but the new profession of engineer was a wholly masculine concern.

4
Collaboration, Reputation and the Business of Mathematics

In 1911 William Henry Young, fellow of Peterhouse College Cambridge, part-time mathematics coach and schools' examiner, applied for the chair of Pure Mathematics at Edinburgh University. His credentials were impressive. As well as experience in education, Young was a research mathematician with three books and ninety-two original papers to his credit, a DSc from Cambridge and expertise in a new field of analysis that was having a profound impact on the development of mathematics on the Continent. Winning the Chair at Edinburgh was important to Young. During the last eight years he had failed to obtain chairs at Kings College London, Liverpool, Durham and Cambridge Universities and, having embarked on research mathematics at the relatively old age of thirty-six (he was now approaching fifty) he was anxious that he should find a suitable post before he got much older. An examiner colleague of Young's, mathematician George M. Minchin of London University, wrote a testimonial emphasising his friend's exceptional reputation and extensive published work. In a private letter to Young he added:

> Perhaps I ought to have said that I was recommending the Firm of W.H. Young & Co. – for I by no means overlook the well-known name of the partner, CGY (Grace Chisholm Young) so well known to mathematicians.[1]

It was in no way remarkable that mathematicians should know Grace's name. Her success in gaining a first-class pass in the mathematics tripos while at Girton College Cambridge, and her subsequent doctorate awarded by the University of Göttingen, had given her some reputation in academic circles. Furthermore, it was claimed that Grace had been the first ever woman to achieve such a distinction in Germany.

That an English woman had succeeded in this way had brought Grace transitory celebrity among the general population and in the national press. Despite Grace's relative prominence in her own time, like so many female mathematicians her achievements have only recently been recovered. Yet, even when remembered by sympathetic scholars, her contributions to mathematics are sometimes relegated to having provided a supporting role to her husband. An account of the Chisholm-Young partnership that has been influential for later authors presents their relationship as a romance between a dutiful, loving wife and a man of (inevitably) disordered genius. She willingly dedicates herself to the role of 'amanuensis' by marshalling his creativity and performing the laborious tasks of writing-up and preparing his offerings for publication:

> … an extraordinary reversal of roles took place … In 1896 the mathematical coach at Peterhouse married the young research mathematician of Girton and Göttingen, each presumably to preserve the same position in the partnership; but now it became clear that the coach had a far more profound and original mathematical mind … and the young research mathematician, the catalyst who caused this profound change, became his secretary and assistant, perfectly capable of making original contributions of her own but basically needed to see that the flood of ideas that was poured out to her could actually be refined into rigorous theorems and results.[2]

In this chapter, the mathematical partnership of Grace Chisholm and William Henry Young will be reassessed with a view to exploring the nature, context and gender politics of their collaboration. The aim is to move beyond simple narratives that all too often assume that individuals are the sole unit of creativity; teamwork cannot be incorporated easily within such romantic visions of the genesis of ideas and, if the partnership is a male-female one, the masculine configuration of 'genius' may often lead to the privileging of the man's role and the undervaluing of the woman.[3]

One description of the Youngs' partnership is provided by Sylvia Wiegand (who is a grand-daughter of the couple).[4] This detailed study presents a thought-provoking account of the Youngs' relationship allowing a more significant role for Grace as a 'creative mathematical thinker'.[5] Despite this, Wiegand is somewhat ambivalent as to whether the two were equal partners, or whether Grace was intellectually subordinate to her husband. Wiegand casts Grace predominantly in the role of 'assistant', describing how 'Grace threw her energy into assisting with Will's research so that he could concentrate on research ideas and "flood the journals"'; and how 'the Youngs' letters to each other give evidence of Grace's assistance to Will'.[6] Yet in conclusion Wiegand remarks that

> On the whole, since each other helped the other's career tremendously and their joint results were far better than the combination of what each could have achieved separately, it seems reasonable to consider them as equal partners.[7]

This ambiguity is a (perhaps unavoidable) reflection of the complexity and varying roles that Grace and her husband adopted within their collaborative relationship, something into which Wiegand offers suggestive insights which have been further developed here. However this chapter also presents evidence to suggest that Grace was not always 'more comfortable in emphasising the family unit over her own identity' and that she and her husband were not always 'satisfied with their personal and professional lives and with the choices they had made'.[8] As other women trying to conform to constrictive ideals of femininity and succeed in a male-dominated field, Grace experienced conflicts which sometimes turned into self-questioning and regret. The case study offered here sheds light on areas that have not been a central question for other accounts. By their very nature case studies can threaten to provide answers which have little application elsewhere. Here this danger has been addressed by the use of other scientific couples to provide context, by consulting memoirs of their children, and by placing the Youngs within the gender mores of their time. Again, the focus is not solely on Grace but on Young's living of a particular kind of masculinity. Additionally, this chapter is contextualised by others which examine the experiences of Hertha and her husband William Ayrton.

The Youngs' relationship over the twenty years in which they were most active was not constant and unchanging but complex and dynamic with each assuming varying roles at different times. Papers could be initiated by Grace, based on mathematics that she had pursued alone; at other times Young took the lead or suggested a problem which husband and wife then worked on together. Grace mentored Young in research mathematics at the beginning of their careers, assisted in the preparation of his lectures, and even instructed him in subjects in which he lacked the specialist expertise to teach. By a detailed look at the correspondence, mathematical notes and autobiographical accounts left by the couple, it will be demonstrated that to consider the Youngs as a 'family firm' is a useful way to characterise the partnership and make Grace's contribution visible. Once decided upon research mathematics as a business, 'the Firm' undertook market research and consulted Grace's eminent mathematical friends as to which area of mathematics would be most profitable. They underwent further training to bring their skills up-to-date and then decided business strategy. In order to secure a prestige academic appointment for Young, most of their product would be marketed under his name. That the vast majority of papers and one of their three books were published under Young's name alone does not imply that they were solely his vision and work. The Youngs used 'W.H. Young' almost like a modern company would use a brand; they felt intense pressure to get their work into the mathematical market place before a competitor got there first and this at times made Young over-exacting in his demands and jealous of any time that Grace spent away from 'firm's business'.

This model of mathematical research recognises that it is a social process in which inspiration and perspiration are inextricably combined. To suggest that Young produced the ideas and Grace the mathematical drudgery is to misrepresent the way their mathematics progressed. As Wiegand suggests, Grace colluded to an extent in losing her mathematical identity within that of her husband's, a finding that will be discussed with reference to other intellectual women who made similar choices. In the concluding section, the Youngs' mathematics will be assessed in relation to their elitist politics and philosophies and it will be suggested that both of these were rooted within the other. The focus of this chapter is 1900–1916, the hectic years during which Grace and Young produced their most important work. But to understand the reasons why their partnership came about requires a consideration of the choices available to the two mathematicians in earlier years.

Mathematical options and personal choices

Grace Emily Chisholm arrived at Girton College at the start of the summer term 1889. She was funded by a small scholarship awarded by the College which was topped-up by an allowance from her father, senior civil servant Henry Chisholm. Grace idolised her father as a highly intelligent, affectionate man who introduced her to quadratic equations and mathematical modelling while her mother 'she supposed, was busy with household duties'.[9] Grace's first paper to the Girton Mathematical Club was on the complexities of her father's Department of Weights and

Measures and she later dedicated her PhD dissertation to him. For Grace, her father represented an elision of rationality and status; these were both the antithesis of her mother's homely concerns and Grace, like others, made an early association between masculinity (her father) and abstract reason. In Virginia Woolf's *Night and Day*, the heroine Katharine Hilbery indulges a secret passion for mathematics as a retreat from the emotional complexities and womanly demands of female life into the contemplation of abstract symbols and geometrical figures.[10] For Grace too, the charm of mathematics resided in its impersonality, necessity, lack of ambiguity and removal from the contingencies of emotional and domestic life. As she wrote as a student at Göttingen to a friend threatening to visit, 'I fancy you would not like to be with me when I am working hard ... I become of necessity quite a different person from the me you know'.[11] In Chapter 1 it was noted that Emily Davies, co-founder of Girton College, encouraged her women to take mathematics as this was (in the first two decades of Girton's existence at least) the most prestigious degree for men. However, that so many of her students complied may well be connected to the subject's aesthetic attractions as a welcome retreat from the narrow prescriptions of femininity around 1900.

Despite her success in the mathematics tripos, the future was uncertain for Grace. For Young, being a wrangler had meant easy transition to a college fellowship and eligibility for mathematical coaching work. For Grace, no such opportunity was a real possibility. As a new college at a time when women's eligibility and aptitude for higher education was still a contested issue, Girton did not have the financial benefactors or resources to fund fellowships and lectureships in the same manner as the older, well-endowed men's colleges. In her personal notebooks, Grace criticises the female dons as being ineffectual and complains that all the main requirements for graduate study – scholarships, fellowships, position at university, the library and laboratories – were closed to women. This was not wholly true, although undoubtedly Grace felt that the obstacles preventing her pursuing graduate research or obtaining a teaching fellowship at Cambridge were overwhelming.

A careers handbook for girls published in 1894 referred to the few teaching posts at the women's colleges at Cambridge and Oxford as poorly paid and chiefly attractive for the pleasant university life that they afforded.[12] For the instruction of 'undergraduates', the women's colleges relied mainly on young fellows from the male colleges and Girton (at least) could afford to pay them only a percentage of the fees that they received for teaching men. Nevertheless, these positions were sought after as Cambridge fellows, since the lifting of the celibacy restriction in 1882, often had wives to support and welcomed the extra income. There is some evidence that Grace was correct in implying that Girton's lecturers were, at times, inexperienced in the academic mores of Cambridge. Helena Swanwick recalled that in 1882 she had tried to engage the Mistress of Girton in discussion about her studies, but she had been less than helpful because 'she knew nothing whatever about the moral sciences (philosophy) tripos'.[13] Nevertheless, in the decades surrounding 1900 several notable female mathematics lecturers did take up teaching posts at Girton, as will be discussed in Chapter 6. However these opportunities were difficult to access, and even after Grace had achieved her doctorate there was little possibility of her being offered a staff position. Still, on her return from Göttingen, there was nothing to be lost from renewing useful acquaintances, so one of Grace's first acts was to send her published dissertation to her old Cambridge mathematical network, including William Henry Young.

In 1895 Grace's future husband was a man desperately seeking a calling that would justify the mathematical expectations placed upon him in his youth. His autobiographical notes suggest that he was tormented by a perceived failure to live up to his school-boy promise by achieving 'great things' at Cambridge – a failure that he blamed on the lack of room for 'creative mathematics' in an examination system geared to fiercely-competitive problem solving. Young had entered Peterhouse College with a scholarship in 1882, been disappointed to be placed only fourth wrangler, yet had gone on to gain a first in the part II examinations the following year. Since then he had been a part-time mathematical coach, investigated entering the legal profession, won a Peterhouse theological prize and engaged in a little experimental physics at the Cavendish Laboratory. Before taking up full-time residence in Cambridge in 1890, he taught at Charterhouse and other public schools and was an examiner at Eton.

Grace first met Young when he became, very briefly, her replacement coach for a few weeks in her last term while her usual coach, Arthur Berry, visited Göttingen. The couple did not conform to the stereotypical relationship of male teacher/mentor and female student however. Young seems to have made little impact on Grace at the time; their daughter recalls that she viewed him as a 'mere boy ... needing help all the time he was supposed to be guiding her' and states conclusively that 'he did *not* tutor her'.[14] Despite this, the 'naturalness' of a woman taking such a subordinate role, and the ease with which this can be extended to a stereotypical model of gendered collaboration, has meant that the myth that Grace was Young's pupil has persisted as a means to characterise their partnership. In a celebration of Young's centenary in 1963, Graham Sutton eulogises the 'great man' but makes only passing reference to his wife as having been one of his pupils, adding (with unfortunate implications) that Young had 'success with less than gifted pupils'.[15] More recently, Wiegand's essay on her grandparents has been placed in the category of 'Couples Beginning in Student-Instructor Relationships' in the edited collection within which it appears. This choice may have been editorial as the author herself gives a far more complex account of the Youngs' collaboration which extends beyond such stereotypes. (In fact, Grace provided more tuition to Young, not the other way around, as will be demonstrated later in this chapter.) Grace had little to do with Young until after her return in triumph from Göttingen in the summer of 1895. Letters began between the two and soon Young was suggesting that they write a book on astronomy together (Grace had taken astronomy as a minor part of her doctorate). The nature of the collaboration that he envisaged soon became clear: he already had a body of notes prepared which required sorting for publication.[16] This project never materialised but correspondence continued, an engagement was arranged, and the couple married in June 1896. Grace was aged twenty-eight years, her husband thirty-two.

Sara Delamont has identified two available lifestyles for the graduates of the new institutions of higher education for women – the 'celibate career woman' and the 'learned wife'.[17] A degree enabled a lady to earn her own living in a new but respectable public role as schoolmistress, headmistress or don, maintaining impeccable moral standards in an all-female community. Although teaching was the preferred choice for the vast majority of female graduates, the women's colleges also had an important function in providing academics with suitable wives. Agnata Ramsey of Girton, who took the only first-class degree in classics in 1887, married the much older and widowed Master of Trinity, Henry Montague Butler. Newnamite Kitty Holt married one of her tutors, Cecil Whetham, and with him co-authored several books on eugenics.[18]

Grace considered herself a mathematician, not a future teacher of mathematics, and fully expected to continue her research career while her husband pursued a conventional life as a Cambridge coach and don. Unlike the physical sciences, Grace needed no laboratory, apprentices or institutional position to engage in mathematics. She already had the qualifications, credibility and contacts. Yet Grace also acknowledged the importance of a family life; as she wrote in a telling phrase to a friend, '… we are not improving the race by letting the best women remain unmarried because they are too exacting'.[19] However, in seeking to be a wife *and* a scholar, Grace was combining two occupations commonly held to be different and incompatible, roles which became increasingly polarised by the 1890s after some two decades of higher education for women. At this time, resentment of women who competed with men became more vociferous and was often felt particularly acutely by male dons' wives. Despite some women students acquiring Cambridge spouses, female scholars and wives increasingly came to be seen as fundamentally different kinds of women. 'Thus the two categories remained ideologically opposed, while some of the actual women involved were induced … to regard one another with little confidence and sometimes little friendliness'.[20] It was tensions of identity such as these that may have contributed to Grace's decision to clothe most of her intellectual product in the name of W.H. Young.

William Henry Young

For Young, Grace must have seemed an attractive proposition. Not only could her reputation in Cambridge mathematical circles be of potential advantage and reflect on him, but her family connections may bring the social prestige that he always perceived that he lacked. (A sensitivity to his parents' status is a subtext in Young's autobiographical jottings.) As a child in Haslemere, Grace had known the Tennysons and William Morris and her father had important government contacts. Shortly after their marriage, Grace lobbied her best friend's father, liberal MP Sir Francis Evans, to pull strings to secure Young's selection as chief examiner in mathematics for Wales. Despite this help, Young failed to secure the appointment. When the couple returned from honeymoon to Cambridge, both found life a little disappointing. Grace was bewildered by the social slights she was receiving from university neighbours and this may have been due to antagonism towards her husband, a prickly man whose self-importance and incipient insecurity combined at times to make him a difficult personality to deal with. In 1919, while attempting to promote an international mathematical tour, Young contacted Harvard professor George Birkhoff in an effort to include his institution in the proposed itinerary. Birkhoff replied rather stiffly 'I am much interested in your analysis of our need for more intellectual cream. My analysis is somewhat different.'[21] Young perceived that Grace might have a useful role to play as a social intermediary; in a letter prior to their marriage he warns her that she must take care that his business letters containing complaints and the like are sufficiently diplomatic.[22]

Young's discontent with Cambridge, and the claustrophobic competitiveness of its politics, can be gleaned from his letters to Grace during his first prolonged absence from her in 1900. He describes himself as a 'fish out of water' and complains that 'the climate here unmans me ... (with) little knowledge and much jealousy'.[23] Young's use of the term 'unmans' is significant. In Chapter 2, the links between manliness and its manifestation in the production of 'great' mathematics which was (crucially) accepted as such by one's mathematical peers, was demonstrated with reference to Göttingen. At Cambridge too, mathematical genius was gendered acutely masculine and Young believed that he was falling short of this ideal. English universities were slow to accept the new, abstract style of mathematics that was represented by Göttingen analysis, and Young smarted at the lack of understanding or respect offered to his chosen specialist field. In addition, he believed that he was not receiving the support or opportunities at Cambridge that should be available to a man of his calibre. This was a time when reforms aimed at bringing mathematical teaching under the auspices of the University meant that demand from men's colleges for coaches was decreasing. With increased competition among coaches for students, fewer women from Newnham and Girton were requiring his services too; in addition he had been sidelined for an examinership which went to Grace's old coach Arthur Berry. Young interpreted his failure to succeed, plus the disempowerment that came with it, as an affront to his manhood. Historians of 'manliness' have emphasised that masculinity is a relational construct which, pre-1914, was informed by a man's reason, social power and privilege over the 'weaker' sex.[24] Young was keen to be independent of the 'whole lot of them' at Cambridge and this would only be achieved by gaining credibility as a successful mathematical researcher and author – something in which his wife was to play a crucial role.

Marriage and the beginnings of mathematical collaboration

The couple began their campaign to get Young's career on track early on in their marriage. Grace wrote to a friend emphasising how 'we must get Will's books written' if he was to achieve academic status and a professorship.[25] Earlier she reported that '... there is so much to do, and however negligent I am of everything except "the book" Will gets in a fever and talks of my many distractions'.[26] This was to be a recurrent theme throughout their partnership and became an even more pressing issue when the children were born, six between 1897 and 1908. The solution was 'girls' to look after the children and Young's unmarried sisters, Ethel and May, to supervise the household. Nevertheless, the pressure on Grace continued and was intense. Fifteen years later, Young's demands on his wife were no less severe with letters admonishing her for writing 'Oh Will, I have too much to do' and urging her to find a suitable girl to prevent the 'whole pack of cards' collapsing.[27] No wonder Grace surmised that Young would have been happy if he had been born Louis Quatorze.[28] Their daughter reminisced that the Young household ran on nervous energy with parents 'too high-pitched' to show affection – that was provided by 'Auntie May'.[29] When in 1900 the family took up permanent residence in Göttingen and Grace decided to consider medicine as a career to 'fall back on', Young veered between support and alarm, urging her not to 'sacrifice your husband and your children to your medicine':

> My chance of getting any sort of position amongst jealous rivals like Berry and Whitehead is in any case small but of course it will be less if you take medicine and are therefore cut off from helping me ... If you decide on maths, and there is much to be said for it, then you will have to work hard at book writing while I am in Cambridge. Will you start on continuous groups say? Whether you decide on medicine or maths, and I naturally hope for my sake it will be the latter, you will have to arrange that the children do not interfere with your work.[30]

Grace acquired a Medical Students' Registration Certificate from London University and began her studies part-time at Göttingen, culminating in a medical diploma in 1904. However mathematics displaced medicine as the 'Firm of W.H. Young & Co.' became securely established. Disenchanted with Cambridge, Grace and Young had left the city in 1897 for a first sojourn in Göttingen, prompted by conversations with Felix Klein, Grace's doctoral supervisor, who had visited Cambridge a few months earlier. Klein's (and Grace's) specialism of geometry was to be the Firm's first product area and in 1898 Young published three papers on this subject in the *Proceedings of the London Mathematical Society*. But to ensure that their mathematical skills were at the leading edge, Grace and Young needed further training. With their young son in tow, they travelled to Italy to study for a few months at Turin University with Professor Corrado Segre, a regular correspondent of Felix Klein, who was developing new ideas in the geometry of higher-dimensional space. They returned to Göttingen in late 1899 where they remained until a move to Geneva in 1908. Young was not a permanent resident abroad; in order to supplement the family's regular income from house leases, savings and investments, he returned for long periods to his job at Cambridge as a part-time coach and examiner. From 1905 Young also earned £100 per year as a special lecturer in mathematics at Liverpool University, six years later winning the title 'Associate Professor'. Despite their constant and prolonged separations, this permanent move to Göttingen heralded the start of the most productive period in Grace and Young's partnership.

Although husbands and male mentors/collaborators have been demonstrated to be one of the main vehicles of entry for women into science,[31] in the case of the Chisholm Youngs it was Grace, not Young, who primarily provided the contacts, secured opportunities for research and influenced the research agenda. As Chapter 6 will describe, women encountered fewer obstacles in infiltrating mathematics as it did not necessarily require an institutional base. Hertha required a laboratory; Grace needed a pencil, a desk and, of course, access to a mathematical education. In addition, the world of pure mathematics, at the top end at least, comprised a relatively small community of individuals who knew (or at least knew *of*) each other. The most prestigious journals – in Germany *Crelle's Journal* and Klein's *Mathematische Annalen*, in England the *Proceedings of the London Mathematical Society* – were read and contributed to by this elite audience, as well as by others. Upon arrival in Göttingen, the couple were immediately a part of Grace's old mathematical network and they both made use of this to further Young's, and the Firm's, interests. Grace persuaded Klein, then one of the most respected names in mathematics, to write a testimonial for her husband. Klein was reluctant initially because, as Grace explained, '... he has hardly ever talked to you about mathematics ... and nearly all our mathematical communications with him have been carried on through me'.[32] Klein also commissioned Young to arrange an English language edition of his Encyclopaedia of Pure Mathematics. Young worked on this at Cambridge but found many of the Cambridge dons ambivalent to the project, in particular about his chief editorship; this prompted the Cambridge University Press to shelve their publishing plans. More significantly, it was Grace again, through her relationship with Klein, who set their business strategy a second time. Klein advised 'the Firm' to read Arthur Schönflies's[33] new work on set theory/functions, intimating that this was an important area of mathematics that was ripe for productive research – and he was right.

Grace and her husband worked to advance this influential branch of mathematics for over two decades.

The theory of sets

Set theory, originating from ideas that Georg Cantor developed in the 1870s, presented what could be called, in Kuhnian terminology, a 'paradigm shift'. As one recent writer has described it: 'set theory opened up a gateway. It was as if you opened a door and on the other side of it was the surface of the Sun.'[34] Sets are at the foundation of modern mathematics because they are central to the way mathematical operations, from subtraction and addition to the logical processes of a computer, are conceived. A set is a group of things, usually having something in common. For example, an apple is in the set of apples, the set of fruit, and the set of objects that are spherical. Sets are also used to define the nature of number. The concept of infinity is intertwined with the theory of sets and it was Cantor who provided mathematicians with the means to use the concept of the infinite to make calculations. Set theory creates paradoxes; it generates multiple infinities (for example, the set of natural numbers is infinite but must be less than the set of real numbers, which includes all numbers including irrationals and transcendentals) and was logically undermined by Bertrand Russell.[35] Yet set theory's power means that, despite all this, it still retains its place at the heart of mathematics. Creating the rules to manipulate sets, creating the axioms (or fundamental assumptions) from which everything that can be known about a set can be deduced, and developing the logic and function theory (a function is a set of rules for turning one number or set into another) to enable mathematical operations, was the contribution of the Youngs to this new discipline.

Although the mathematical tools developed by Grace and her husband would be applied to physical problems (such as calculating velocity and problems in astronomy) their actual work was highly abstract and removed from particular everyday problems. William Henry Young's name is attached to the discovery, independently but around the same time as Henri Lebesgue, of the 'Lebesgue integral'. This solves the problem of defining the properties of shapes by reducing them to a set of points. The Youngs' 1906 book had as its title *The Theory of Sets of Points* and is seen as fundamental to introducing complex function theory to Cambridge. Through her experience and contact with Klein and his Göttingen school, Grace was instrumental in setting this research programme; she was also instrumental in its execution. The couple's extensive correspondence builds a vivid picture of the ways in which they collaborated. Grace's role can be identified as maintaining 'the Firm's' up-to-date mathematical expertise from her base in Göttingen; engaging in 'mathematical networking'; using her greater technical skills to hone papers and ensure accurate proofs; contributing ideas and research developed alone; and writing up the papers (her own, her husband's and papers published under both names). Grace also supported Young in his teaching by researching lecture material, assisting in marking, and providing him with coaching when he was required to teach subjects beyond his experience.

The Youngs' partnership unravelled

In 1902, Young writes from Cambridge of his annoyance that he has not been asked to examine for part II of the tripos because he 'doesn't know functions' and asks Grace to explain Schönflies's oscillating functions to him.[36] This was a typical pattern for the couple from around 1900 to 1908. Grace, based in Göttingen where research was being originated, was able to ensure that Young stayed up to date with developments even though he was far away in Cambridge. She attended regular 'colloquiums' or weekly seminars with the professors at her old university and frequently reported the mathematical news and views of Klein, Hilbert and the others with whom she maintained social, as well as academic, relations. This was performed knowingly and was not merely incidental. Grace sends Young her notes from these events and responds to her husband's urging to obtain from Klein, and critique herself, pre-publication proofs of other mathematicians' work (including Schönflies's). In 1903 she copies out a paper of Henri Lebesgue for him, warning that it is 'hard to understand', and sends him an example of a theorem 'which you and I have been hammering at for so long'.[37] Grace also 'sounds out' Klein about possible papers for Young to publish: 'The impression left on my mind was that a good lucid exposition of the subject would meet with his approval, though you understand, I hinted at nothing.[38]

There is no doubt that 'the Firm' relied on Grace to maintain high levels of quality assurance for their product. Their correspondence is littered with requests to Grace to check his (and his rivals') reasoning and find flaws, and Young sends her copies of theorems that he is having difficulty proving. In 1903 Young writes of a stubborn problem on the upper integral and his pleasure that Grace can work it out further for him (his paper on the subject appeared the following year).[39]

As late as 1914 Grace points out to her husband that a theorem of his published in the *Proceedings of the Cambridge Philosophical Society* in 1910 was untenable, noting that at the time they had both believed that it had 'come out too easily'. She then outlines the fallacy and rectifies the problem.[40] Grace did not take a passive role in the mathematical partnership and would often take the lead, for example by withdrawing papers when Young wanted to publish too early. In January 1906 Grace is peremptory in ordering her husband to withdraw one of his papers due to errors:

> The fact is the definition won't do: with your definition, as it stands, we do not necessarily get, in the case of an ordinary function defined for a segment, the 'upper integral' in the old sense, at all. Secondly, with your definition, as it stands, the theorem upper integral = upper integral of associated upper is not true. See the following example ...[41]

The combined contributions of Grace and her husband blur any easy distinction between 'ideas' and 'technical execution'. There are numerous examples of Grace suggesting the subject matter of a paper and her diary entries show how she worked hard at developing new techniques and not just solving problems. In 1912 Grace writes of grappling to find a new method to apply to Fourier series, noting '... working at multiple Fourier series all day, got it all finished by 11.30pm'.[42] This was a subject upon which Young published several well-received papers in the *Proceedings of the London Mathematical Society* between 1910 and 1916.

Grace has been credited with the 'writing-out' of her husband's papers, turning his creative ideas into lucid text. Certainly Grace contributed enormously to the intelligibility of Young's papers and managed all the administrative labours connected with proofreading and publication. When Grace ceased work in despair at the death of her eldest son in 1917, letters begin appearing from G.H. Hardy, Young's main referee for the *London Mathematical Society*, questioning Young's 'diffuse' writing and 'untidy, inconsistent referencing'.[43] However the couple also worked the other way around and it is too simplistic, given the high level of expertise of both, to assume that their roles did not change and overlap. For example, at times Young sends Grace problems for which he needs solutions on the understanding that, once he receives her answers, he will write out the papers himself.

Young held various teaching and examining positions during this time and relied on Grace to provide support for this too. In 1901 she marked Cambridge 'Little Go' papers and planned a course on geometry for her husband. When Young was finally appointed to the Central Welsh Examinations Board in 1902–5, Grace translated past papers into German and arranged for her German mathematical friends to evaluate them. When in 1913 he was appointed to an Associate Professorship in the History and Philosophy of Mathematics at Liverpool University, Grace commenced studies in philosophy, corresponded with Bertrand Russell, and reported to her husband that she had 'matter enough for a whole course of lectures for you'.[44] Given the extent of Grace's contribution to the 'Firm of W.H. Young & Co.', it is difficult to see how her role could be reduced to that of mere secretary and assistant. Certainly in one letter to her mother Grace writes of how exciting her husband's papers are and how she helps him 'as secretary and critic a little'.[45] However Grace was keen to build up Young's prestige to a family who did not like him and who were hostile to the couple's move away from England. Grace's marriage heralded a rift with her parents, especially her mother, which lasted until they died in the early 1900s. Simultaneously as she wrote to her mother, Grace takes joint ownership of her husband's papers in a letter to a close friend by explaining that 'We are getting on with the mathematics, have finished the second paper, which was hard work, and have got a good deal of the third down'.[46] Assigning to Grace the role of directed labour, performing the routine mathematical tasks that freed Young to use his genius unencumbered by detail, is in keeping with a gendered assignment of duties, reflected in later historiography. Recent scholars have shown how women were inevitably assigned the role of 'assistant' in both-sex collaborations, whatever the nature of their individual contributions.[47]

Another celebrated mathematical research partnership, active at around the same time as the Youngs, is that of Oxbridge professors G.H. Hardy and J.E. Littlewood. Hardy and Littlewood, who collaborated for thirty-five years from 1911, were analysts like the Youngs and have been credited with creating a school of mathematical analysis at Cambridge.[48] In historical accounts of their work a similar division of roles is assigned to their male partnership as has been assigned to the Youngs:

> (Hardy's) ... long collaboration with Littlewood produced mathematics of the highest quality. It was a collaboration in which Hardy acknowledged Littlewood's greater technical mathematical skills,

but at the same time Hardy brought great talents of mathematical insight and a great ability to write their work up in papers with great clarity.[49]

Despite this division of labour, the Hardy-Littlewood partnership is rarely forced into conformity with a model of 'genius and assistant', quite the reverse. A recent work of scholarship describes their collaboration as the 'most remarkable and successful partnership in mathematical history ... these giants produced around one hundred joint papers ...'. Littlewood's role is described as making the 'logical skeleton, in shorthand' while Hardy's contribution was to provide proofs and write up joint papers. Despite this division of labour they have *both* been characterised as mathematical greats. In this account, far from 'writing up' being presented as a mundane, derivative task, when Hardy performs the role he is represented as 'the consummate craftsman, a connoisseur of beautiful mathematical patterns and a master of stylish writing'.[50] In this all-male partnership, the 'technical skills' of one collaborator have not been used to downgrade his contribution to work produced jointly; 'technical' aptitude and a flair for 'writing up' tends to be undervalued only when a gender dimension is added. The above description of the Hardy-Littlewood collaboration, as well as the example of the Youngs, problematises perceptions of mathematical research that rigidly distinguish between 'ideas' and 'technique' and assign the former to individual intellects alone. It is difficult to separate the two or to give one a privileged position in the development of mathematics. Collaboration is common in mathematical research, despite the persistence of 'lone-genius' mythology, and Hardy and Littlewood both had other collaborators besides each other.

Of course, Hardy and Littlewood published their papers under joint names; this was not often the case with Grace and her husband. The publishing strategy of the 'Firm of W.H. Young & Co.' was to achieve prestige and position for Young, and they needed them quickly to secure their financial situation. It would have been difficult to earn a living solely by writing, advances were rare and most contracts stipulated a fixed fee for receipt of work. Rewards were not high and most male authors combined writing with another career. When Young attempted to publish Klein's encyclopaedia in English he had to approach the London Mathematical Society and the Royal Society for funding; when the Cambridge University Press finally turned down the project, Macmillan Publishers suggested trial runs of small volumes retailing at just one shilling each. Grace published only four papers under her name alone (including her PhD dissertation) up to 1905; she resumed in earnest in 1914[51] after Young was appointed to an associate chair at Liverpool University and, shortly after, Calcutta University. Grace's later work included an important series of papers on the foundations of differential calculus. Up to 1914 Young published approximately eighty-eight papers. Six papers were published jointly and it is difficult to see why Grace and her husband broke their publishing routine in these cases. Possibly one was published in both names because it was for an Italian journal, the others because Grace had done the majority of the research. Three books appeared during this time too, two (*A First Book of Geometry*, 1905, and *The Theory of Sets of Points*, 1906) were published in joint names, the other (*The Fundamental Theorems of the Differential Calculus*, 1910) in Young's name alone. It is possible that *A First Book of Geometry* was written solely by Grace. It was published in her name alone in the German edition and correspondence suggests that it was indeed her project. Certainly this was a basic school textbook and therefore within the remit of women authors who were channelled into writing for children or educated laypeople. When publishing for a more scholarly audience, it was harder for women to establish their credibility. Grace describes in letters and her diary how she designed paper models and drawings for the geometry book with her young son; this was an activity which she had loved to do with her father as a child.

Division of 'the laurels'

In a 1902 letter marking the start of the most productive years of their collaboration, Young outlines his programme for the couple and their publishing aims:

> I hope you enjoy this working for me. On the whole I think it is at present at any rate quite as it should be, seeing that we are responsible only to ourselves as to the division of the laurels. The work is not of a character to cause conflicting claims. I am very happy that you are getting on with the ideas. I feel partly as if I were teaching you, and setting you problems which I could not quite do myself but would coach you to. Then again, I think of myself as like Klein furnishing the steam required – the initiative, the guidance. But I feel confident too that we are rising <u>together</u> to new heights. You do need a good deal of criticism when you are at your best, and in your best working vein. The fact is that our papers ought to be published under our joint names, but if this were done neither of us get the

benefit of it. No. Mine the laurels now and the knowledge. Yours the knowledge only. Do you suppose people will venture to say the laurels ought to be yours? No, they would be very unwilling to allow that … divide and we are lost. Everything under my name now, and later when the loaves and fishes are no more procurable in that way, everything, or much, under your name. There is my programme. At present you cannot undertake a public career. You have your children.[52]

Young's words – *Do you suppose people will venture to say the laurels ought to be yours? No, they would be very unwilling to allow that …* – seem to be anticipating a protest by Grace, and could even be read as a warning to her. Is her husband asserting that people would not accept that a woman had done the work? More likely, Young is implying that people would frown on a wife who took credit away from her husband; antipathy to women's intellect was especially strong when it rivalled *publicly* the intellect of men. This was one of the arguments put forward by opponents of higher education for women who urged complementarity between the sexes, fearing that 'unnatural' competition would be the result if men and women were to operate in the same sphere. This stricture was even more compelling when the individuals concerned were man and wife, as wifely duties required being a support to one's husband, not a rival. In fact, being a rival to one's husband could be interpreted as competing against yourself: a man and his wife were regarded as 'one' by sections of society opinion and by the law[53] and this could impact on women seeking an independent identity. When Hertha Ayrton sought election to the Royal Society this was refused (nominally at least) on the grounds that as a married woman her person was included in that of her husband and, therefore, she was ineligible for consideration (her husband, William Ayrton, was already a fellow).

Although, to a large extent, Grace acquiesced in absorbing her mathematical identity into that of her husband's, at times she also rebelled against it. Despite Young's admission that papers should really be published under both their names, in practice he was very careful, in any public context at least, to exert his superiority. For example, in the preface to their 1906 book, to which Grace contributed at least equally to her husband, Young writes

Any reference to the constant assistance which I have received during my work from my wife is superfluous, since, with the consent of the

Syndics of the Press, her name has been associated with mine on the title-page.[54]

Young hoped that a good reception for his book would lead to him being offered a professorship (which was what they both wanted and needed) and he did not want his recognition diluted by being shared with his wife – a woman who had already acquired a mathematical reputation and who was well-known in mathematical circles.[55] The same concerns are discernible on the pages of the *Proceedings of the London Mathematical Society* which was an important outlet for the Youngs' papers. Young is careful to maintain his dominance in this mathematical journal; for example, in the volume in which Grace's name first appears as co-author of a joint paper with her husband, Young also publishes four papers under his name alone. This was typical and Young was a highly-active contributor to the *Proceedings* in the first decade of the twentieth century, publishing up to seven papers each year (while simultaneously holding down a full-time job). Such was Young's productivity during this time that it is noted in more than one memoir as exceptional and in need of explanation. For Sutton, Young's life is a 'great enigma': it is 'an extraordinary thing that a man should spend so much time on the hack work of coaching and then suddenly spring into prominence as one of the most prolific and greatest analysts of his time'.[56] Like Sutton, Hardy refers to Young's late start but notes 'the productivity, when it did come, was so astonishing' and adds that 'the dreams were to come and the "drudgery" to end, and the end came quickly after Young's marriage'.[57]

There is no doubt that Young benefited the more from their collaboration. He was awarded a DSc from Cambridge in 1907 on the basis of his published papers and in the same year was elected a fellow of the Royal Society. In 1913 he was appointed Associate Professor at Liverpool University, a few months later the first Hardinge Professor of Mathematics at Calcutta University, and in 1919 Professor of Pure Mathematics at the University of Wales at Aberystwyth. The London Mathematical Society awarded him the De Morgan Medal in 1917 and the Royal Society awarded him its Sylvester Medal in 1928; in the same year he was elected an honorary doctor at Strasbourg University. Young served as President of the London Mathematical Society 1922–24 and in 1929 was appointed President of the International Mathematical Union. Three years before he died he was elected to an honorary fellowship at his old Cambridge College, Peterhouse. Grace and Young's story highlights the futility of looking to male frameworks of achievement in order to establish women's participation in mathematics or science. An examination of published papers, scientific awards and prestigious appointments will throw light only on Young's career; Grace's achievements will remain in the shade. To render women's contributions visible it is important to look beyond the public sphere. As one historian has suggested, the unbalanced gender structure of modern science may result 'not so much from the exclusion of women from science, but rather from the exclusion of the *domestic* realm from science, and the incidental concomitant exclusion of women'.[58]

Intellectual women and 'womanly' roles

Grace has been represented as a dutiful, self-sacrificing wife who willingly used her mathematical skills to aid her husband and further his career. Although this is partly true, the complete picture is more complex. Certainly Grace felt that she and her husband were working in partnership for the firm of 'W.H. Young & Co.'. She is quoted as preferring to remain 'incognito to the outside world ... husband and wife being one' and fearful of being seen as 'ambitious for herself and her own glorification'.[59] Indifferent to women's suffrage, part of her adhered to a Ruskinesque ideal of women as supporters and enablers of men that was an influential belief which persisted into the first years of the twentieth century. Grace's academic background may have contributed to, rather than questioned, such ideas. Carol Dyhouse has suggested that in some ways the new educational institutions like Girton functioned as conservative establishments fostering conventional values and ideals about femininity and female service.[60] A common allusion in the Youngs' correspondence is an admiration for the Brownings; Grace was an admirer from childhood of Elizabeth Barrett Browning's novel-poem, *Aurora Leigh*. This epic was very popular and had passed through five editions by the time Barrett Browning died in 1861. Deirdre David has demonstrated how, within the poem, women's intellectual product is made the servitor of male cultural authority, reflecting Barrett Browning's belief in the ultimate superiority of men's minds and the 'yearning' that women have to 'lose ourselves' in male partners. David's conclusion that in the poem we hear a woman's voice speaking patriarchal discourse – boldly, passionately and without rancour – is to some extent descriptive of Grace.[61]

Grace's autobiography is a romanticised account of her Girton years along the lines of L.T. Meade's popular novels of the time about the fictitious women's college 'Merton'.[62] Written with hindsight, in the third person and with names changed, Grace portrays herself as a tall, slender Girtonian called Iris while Young is turned into a 'Darcy-like' hero called Mr. King whose ridicule reduces his female students to tears.[63] To use this material uncritically is to ignore the tensions that clearly emerge in Grace's letters, diaries and notes. Despite acquiescence to her husband, Grace's desire for recognition did not always lie quiet and Young was not above pressing his claims on her time and intellectual product. In one of Grace's notebooks she pens some personal thoughts amongst the mathematical jottings, including her concern that Young says people 'despise her' plus this silent 'answering back':

> Will said to me today 'A woman ought not to mind playing second fiddle, it is not really such a hardship. Women play second fiddle far more satisfactorily. Think of Mrs Browning. If instead of working in separate rooms at their separate poems, she had thrown herself with Browning's work and helped him to express himself more clearly and exactly, and think more conscientiously, his work would have been infinitely better, and a great deal of her stuff would never have been published. She may have published two or three really good things herself and nothing lost.[64]

Grace suffered a particular bout of depression around New Year 1900 when she wrote, in tears to a close friend, that all of her hopes for her future, by which she meant her mathematical ambitions, were disappearing.[65] Her friend, who had been a fellow student at Girton, emphasised ideals of female service in her reply and advised that she could not think mathematics 'a duty higher than one's duty to one's family'.[66] It is significant that as soon as her husband had secured a worthwhile position Grace resumed her own publications.

For most of their married life, Young was absent from the home for long periods of time. He maintained rooms at Peterhouse College and so remained to a large extent in a masculine, manly environment where femininity rarely infiltrated. To colleagues he implied that he preferred to be separated from his family during the working period.[67] Nevertheless, he liked to assert his authority over the household and insisted on being informed of the smallest detail, controlled the children's daily schedules and even obliged Grace to send him her accounts. In 1871 Samuel Smiles wrote that it is in the home that 'a man's real character ... his manliness is displayed'. Young can be aligned within a later-Victorian concept of manhood in which masculinity is never fully achieved but has to be constantly asserted and renegotiated. John Tosh has warned that the notion of 'separate spheres' can be misleading in suggesting a true complementarity in which each spouse is sovereign in his or her domain. He suggests that men, especially those engaged in sedentary, 'feminised' intellectual work, required a forceful assertion of will over wife and home to realise their masculinity.[68] Unlike Hertha's husband, who was secure in the gendered, masculine codes of his profession, Young may have felt the need to assert his masculinity more explicitly. After the estrangement and later death of Grace's father in 1901, Young took to himself the roles of both husband and father.[69] This double dose of Victorian patriarchy dominated Grace in her role as mathematician, as well as in her roles as wife – and 'daughter'.

In the late Victorian/Edwardian era literary images of scientific and learned women were used in books and girls' periodicals to instil ideals of service and teach moral lessons; this construction of scientific women as 'selfless heroines' helped preserve gender hierarchies within a science – and a world – in which women were asking for greater participation. Mathematician Mary Somerville was commonly represented as an ideal of learned femininity for selflessly dedicating her life to bringing the astronomical writings of Laplace to an English audience.[70] Eleanor Sidgwick was represented and praised by her niece for having 'silently renounced' study for the Cambridge mathematics tripos in favour of supporting her husband and his work.[71] Just such an example was set by the short story, 'An Admirable Arrangement', which appeared in an 1897 edition of *Lady's Realm*. A Cambridge don is dismayed that the Girton graduate who is fellow guest at a house party has done original research into the primitive tribe which is his specialist area; what's more she has produced a brilliant new theory to rival seriously his own, jeopardising his academic reputation. The narrative is resolved by the two falling in love, with the girl promising to marry the don on condition that he include her work in his next book, putting only his name on the cover.[72]

Grace can be aligned, in part, with other intellectual women who absorbed this moral message, shunned public recognition, and submerged their talents to further the interests of male relations. Margaret Huggins took a pivotal role in research with her celebrated husband, astronomer and President of the Royal Society William Huggins, whom she married in 1871. Margaret brought her own photographic expertise to the partnership, developed new techniques of research and carried out much of it herself. Despite this, Barbara Becker has demonstrated how the couple colluded to present a traditional and romanticised image of themselves with William as the principal investigator and Margaret as his able assistant.[73] Barbara Caine in her book on Beatrice Webb and her sisters reaches similar conclusions as to the limited nature of many women's ambitions, despite the changes of the late nineteenth century. For the 'Potter girls' public activity provided a way of supporting and assisting husbands; even Beatrice who, like Grace, had already made a mark in her chosen path, dedicated two thirds of her career while married to furthering her husband's work and political ambitions.[74] Sara Burstall, who remained single, recognised this dynamic and countered the accusation that women had not produced a high enough level of intellectual work to justify their admission to university by arguing that the chief reason for this was that so many of the ablest women marry and help their husband in his research instead.[75]

Conclusion

Despite her partial acquiescence to submerging her mathematical identity with that of her husband, Grace can be interpreted as a 'feminist'[76] although she did not support the suffrage cause and spoke against it at a college debate in her first year at Girton. Recent scholarship on 'the new woman', commonly linked with 'graduates' from the new women's colleges, has problematised the one-dimensional portrayal of this figure as one who privileged independence over family and rejected sexual difference or differing social roles based on biology.[77] Eugenic feminism was a strong thread in Edwardian Britain, predicated on woman's role as mother, and Grace exhibited strong sympathies with this; she admired the novels of Sarah Grand[78] and, as outlined in Chapter 2, wrote a personal handbook for girls and their guardians which explicitly endorsed such views.

The Youngs' mathematics also informed, and was in turn informed by, an elitist politics that developed in later years to criticise wider access to education as 'rewarding memory at the expense of intellect' and to describe socialism as a 'monstrous doctrine promulgated by the commercial classes'. This flight to exclusive extremism was made with direct reference to the growth of the practical sciences and their (relative) democratisation of access (to men) based on acquired skill and meritocratic principles. In an unpublished article headed 'Bureaucracy and Intellect at Cambridge and the Royal Society', Young rails against the devaluation of 'intellect' at these institutions and argues that within 'our modern civilisation ... every science is one-sided to the extreme, so much so as to be almost vulgar. Each science has come to be the occupation of inferior minds'. Again, the emphasis is on individual minds – special intellects with which the natural elite are born and which can be acquired in no other way. Despite this, in the wake of World War One both Grace and her husband were committed to fostering international connection and co-operation in mathematics and Young was instrumental in the establishment of the International Mathematical Union (IMU); he served as the first IMU vice-president in 1919 and president 1929–1932. Yet in the 1930s the couple flirted with fascism as 'competition between races becomes inevitable when there are inferior types' and Young entered into correspondence with Mussolini and Oswald Mosely.[79]

To recognise Grace's mathematical contributions is not to downgrade her husband's achievements; an exploration of the tensions within their partnership is not a denial of the existence of genuine affection and love between the two. To see beyond the standard narrative of strong, intellectual manhood and selfless, secondary womanhood, is a start in unlocking the complexities that these hierarchies hide. It is only in so doing that the major misrepresentation and under representation of women in mathematics and science can be fully understood. Yet, as illustrated in the previous chapter, any woman seeking to step outside this narrative confronted obstacles that women who negotiated their mathematical or scientific identities within the stereotypical ideals of 'feminine service' did not meet. Hertha repudiated these stereotypes, argued for equality between the sexes and refused to be placed on a pedestal. Grace came to believe in a special role for women and predicated her feminism on the 'difference' of the female sex. Yet, whether they embraced 'new' or 'traditional' attitudes to a 'woman's place', they both encountered problems in reconciling their femininity with recognition for their intellectual pursuits. In the case of experimental science, this will be illustrated in the next chapter with reference to Hertha Ayrton and her struggle to achieve parity in the laboratory.

5
The Laboratory: A Suitable Place for a Woman? Women, Masculinity and Laboratory Culture

In his inaugural lecture in 1871, James Clerk Maxwell, recently appointed Professor to the soon to be opened Cavendish Laboratory at Cambridge University, stressed that his new facility's prime focus would be experimentation for illustration and for research. One of his first projects was to repeat the experiments of the 'great man' to whom the new laboratory was dedicated. In Henry Cavendish's day (d. 1810) there was, as yet, no instrumentation with which to measure an electric current, so Cavendish passed the current through his own body and estimated its magnitude by the intensity of the resulting shock. In keeping with the experimental bravery of his facility's namesake, Maxwell set up similar apparatus at the new Cavendish Laboratory

> ... and all visitors were required to submit themselves to the ordeal of impersonating a galvanometer. On one occasion a young American astronomer expressed his severe disappointment that after travelling to Cambridge on purpose to meet Maxwell and consult him on some astronomical topic he was almost compelled to take off his coat, plunge his hands into basins of water and submit himself to a series of electrical shocks![1]

Not surprisingly, given Maxwell's encouragement of a culture of physical courage and stoicism, women at Cambridge were not welcomed at the Cavendish Laboratory during his tenure. It was only in 1882, during the professorship of his successor, Lord Rayleigh, that women were admitted on the same terms as men.[2]

An analysis of obituaries, memoirs and biographies of male scientists active in the decades surrounding 1900 presents a striking illustration of the way in which the laboratory was presented as a masculine space where heroic qualities could be tested, developed and displayed. So a reminiscence of James Dewar, who experimented on the liquefaction of gases in the early 1900s, highlights the 'personal courage' and 'iron nerve' that his work required. When 'an alarming explosion rent the air of the laboratory' Dewar 'did not move a muscle, or even turn to look'. This memoir continues:

> Dewar never admitted that anything was dangerous. The most he would say was that it was a little tricky. Considering that Lennox and Heath, his two assistants, each lost an eye in the course of the work, this was certainly not an overstatement.[3]

Such daring in the laboratory and the forbearance of uncomfortable physical conditions are common in the recollections of nineteenth-century men of science. Even if some of these memoirs are apocryphal, that the laboratory was represented as a site of manly display and the development of moral and physical courage, as well as of the production of knowledge, is significant when considering the experiences of women in science and the representation (or lack of it) of women in the laboratory.[4]

Until recently, women's absence from the laboratory has been explained largely by the growing institutionalisation of science at the end of the nineteenth century which acted as a barrier to women's participation. As the tradition of home laboratories gave way to new, specialised, experimental facilities, so women were marooned in the domestic sphere from which science had fled.[5] Recent scholarship has revealed a more complicated picture by uncovering previously 'invisible' women and following them into the laboratory.[6] Responding to this new visibility of women, this chapter suggests that the new professional scientist's need for moral and material status necessitated the representation of a heroic laboratory culture which was antithetical to femininity and, by necessity, ignored female experimenters. This discourse worked alongside the forces of professionalisation and institutionalisation – forces which require unpackaging to reveal the complex mechanisms of inclusion and exclusion.

The term 'laboratory culture' is used here in a broad sense to encompass both the shared experiences of workers in the laboratory and the way in which these experiences were represented to a wider public. The primary focus will be on physical laboratories, as these were where Hertha sought to work and where the representation of a heroic, manly culture reached its highest potential. Biological, chemical and other specialist laboratories present with their own histories, culture and aesthetics which necessitate further, individual analysis.[7] Anecdotes of laboratory life developed into a mythology which was picked up by journalism and fiction (which in turn reflected it back to the laboratory) and which was shared by both scientific and non-scientific audiences alike. To illustrate this, it is necessary to adopt an inclusive approach which goes beyond the testimony of scientists and scientific writing to include literary, fictional and photographic sources too. In this way, the processes that worked to affect the experiences, expectations and representations of Hertha and other woman active in experimental, laboratory-based science begin to be uncovered.

The growth and significance of laboratories

The closing decades of the nineteenth century are characterised by historians of science as the years of the 'laboratory revolution'. Between 1880 and 1914 there was an enormous growth in institutional laboratories as the older 'devotee' tradition of research undertaken by gentlemen at home gave way to a new professionalism, and as the universities embraced the newer natural sciences that required experimental facilities. By the end of the century laboratories were used for teaching, research and commercial purposes, plus they could function as workshops and places of production as well as of discovery. Amid fears that Britain was trailing behind Germany in technical innovation, well-equipped laboratories and workshops became symbolic of national well-being and a key instrument in the race for international competitiveness. In the universities, possession of experimental facilities was becoming increasingly important and women's colleges were not left out of the expansion. At Cambridge, chemical laboratories were built at Newnham and Girton Colleges in 1879, and the Balfour Laboratory for Life Sciences for Women at Cambridge was opened in 1884. Towards the end of the century Bedford College, London, possessed six new laboratories, erected at a cost of more than £6000, and Royal Holloway College, opened by Queen Victoria in 1886, boasted well-equipped chemical and biological laboratories. The former prompted *Nature* to remark that it was 'surely a hopeful sign that a college for the education of women should now be regarded as incomplete unless it contains physical and chemical laboratories specially designed and fitted for the delivery of lectures and performance of experiments'.[8] However, that scientific women in universities were largely catered for with parallel facilities limited their impact on the culture that grew up around laboratory experimentation; this gender separation also suggests a disquiet about women sharing the laboratory space with men.

In the limited instances where laboratory facilities were shared, women were a small minority. Only a handful of women have been identified as 'researchers' at the Cavendish Laboratory in the nineteenth century[9] and, as discussed in Chapter 3, Hertha was one of only three women amongst 118 men when she embarked on an evening course in electrical and applied physics at Finsbury Technical College in 1884.[10] Even when the emphasis was on training science teachers rather than on technology and industry, women's participation was negligible. In 1896–7 seasonal courses offered by the Royal College of Science at South Kensington (the Central's partner institution) were attended by 300 men but only six women. The previous year's short summer courses, with an emphasis on practical laboratory work, had attracted 202 male and ten female students. The compiler of these figures, writing in 1897, placed the blame for this under-representation on women themselves for failing to take advantage of the academic opportunities placed within their reach.[11] This is a criticism not unfamiliar to modern ears and one which suggests a more subtle causation of women's absence. Historians of women and science have demonstrated the ways in which the growing institutionalisation of science worked to keep women out, however the cultural meaning of the laboratory and its effects on women have been largely overlooked. The laboratory's moral currency as a venue in which to develop and exhibit manly character, plus the increasing significance to national pride and virility of success in this experimental, knowledge-and-technology producing space, were all elements which interwove to keep women to the periphery. Even the tradition of home-based laboratories, which did not disappear entirely but co-existed with institutional laboratories, did little to render women more visible in science, as Hertha's career illustrates.

The laboratory as a site of manly display

The tendency to represent the laboratory as a site of manly display, heroic endeavour and moral bravery created an inevitable ambivalence towards femininity. Writers of scientific memoirs used experimental work as a vehicle for constructing personas that recall the adventurer-heroes of Edwardian fiction such as Conrad's Marlow (Heart of Darkness) or H. Rider Haggard's Allan Quatermain (King Solomon's Mines).[12] Speaking of the Cavendish Laboratory, one researcher remembered being a near-victim to wires so arranged overhead that they threatened passers by with decapitation, and recalled battery cells containing nitric and sulphuric acids which 'what with the fumes which assailed one's throat and the acid which destroyed one's clothes' were 'a most disagreeable business'.[13] One British visitor to Edison's laboratory in the States likened the great man to Napoleon the First and affirmed that 'All one hears about his working three or four days and nights at a stretch is quite true: he has about 100 assistants, and manages, I hear, to keep them working all night too very often'.[14] It was 'family legend' in Hertha and William Ayrton's household that while running a teaching laboratory in Tokyo, Ayrton 'after two suicides and a murder ... discouraged ritual sword-wielding by discharging a large revolver into the ceiling of his small laboratory'.[15]

A photograph of Rayleigh's laboratory at the Royal Institution, where he performed work leading to the discovery of argon, has captured the culture of the physical laboratory by giving emphasis to a sign hanging from the ceiling that warns simply 'DANGER' (Figure 5.1). In some accounts, the harsh physical environment of the laboratory is heightened by competitiveness, mainly for scarce equipment, redolent of the public school playing field. The 'danger' of Rayleigh's laboratory was augmented by its laboratory store being 'regarded as a plundering ground by the Scottish marauders from downstairs' (the chemists in the basement).[16] In the face of these trials and challenges, the scientist in his laboratory was objective and calm under pressure, doing his job just as professionally as the colonial administrator in similarly difficult and potentially dangerous places. In this way, the laboratory researcher's identity managed to combine the rationalism of scientific methodology with the heroic potential of experimental work.

Unease with how a female scientific experimenter should be presented is reflected in the nature and scarcity of fictional and journalistic representations of the day. It is rare to find a woman pictured as an active researcher and experimenter (as opposed to a student) in the laboratory. Science and scientists were a favourite subject for the rapidly expanding Edwardian periodical and magazine press yet a study of media science before 1914 mentions no women scientists as subjects of reports (although there were a significant number of women science writers: women were accepted as popularisers of science for an amateur or child audience and could call on a tradition of such female authors to justify their work).[17] Ideas concerning the biological and emotional unsuitability of women for dispassionate intellectual labour and the dangers to health and fertility that might accompany attempts at such manly pursuits (loss

Figure 5.1   Physical Laboratory, Royal Institution, towards the end of the nineteenth century. Images from Lord Rayleigh, Robert John Strutt, 'Some reminiscences of scientific workers of the past generation, and their surroundings', *Proceedings of the Physical Society*, 48 (2) (1936)

of beauty, fertility and womanhood) were a theme that added to a nervousness of representing women as scientists or experimenters within the same frames of reference as were applied to men.

The valorisation of scientific heroes in the periodical press was contrasted by a common fictional portrayal of scientists as male, unemotional, detached 'and, at worse, inhuman and insane'.[18] This was the ideal of the 'objective' scientist pushed to extremes, an example of masculinity in direct opposition to passionate, empathetic femininity. The decades of the laboratory revolution were also when the new genre of 'scientific romances' (H.G. Wells, Jules Verne, Robert Louis Stevenson) became popular. Much of the fast-growing output of magazine fiction also used science or scientific ideas as a favourite theme with stories written following the tradition of Faust, Frankenstein and the gothic novel. Here we can see laboratory culture and popular culture spilling over and informing each other. It has been noted that in these texts it is often the laboratory which has 'diabolical connotations' and is the site of danger and evil.[19] It is possible that these representations had associations with the growth of a descriptive language which emphasised courage and manliness in the experimental space. The laboratory signified a place of unease in some parts of popular literary culture and, as these stories often stood alongside 'factual' accounts of science and scientists, it can be assumed that science and science fiction had a reciprocal effect upon each other.

The stereotype of the laboratory as a dangerous place calling for physical and psychological bravery and endurance by necessity omits any feminisation or representation of women. It is difficult to find evidence of the experiences of women who worked in this environment; no wonder there is little scholarship in this area. Marsha Richmond has written on the history of the Balfour Biological Laboratory for Women at Cambridge University and notes an undercurrent of prejudice against the few female research workers (as opposed to undergraduate students from Newnham and Girton who attended lectures and demonstrations there) which 'could sometimes turn into hostility and make working there problematic'. She quotes a male researcher talking about Cambridge in the 1890s:

At that time women were rare in scientific laboratories and their presence was by no means generally acceptable – indeed that is too mild a phrase. Those whose memories go back so far will recollect how unacceptability not infrequently flamed into hostility.[20]

Although laboratories were built at some women's colleges in the first decade of the twentieth century, as previously outlined, these were mostly teaching, not research, institutions. At Cambridge, their function was to offer routine instruction to female undergraduates who were segregated in their own facilities and did not (as a rule) have opportunity to undertake novel experimental work. As a result, these scientific women did not pose a threat to the reproduction of manly values within University-wide facilities which owed their reputation to research. As late as 1911, William Ramsay at University College London was reported to 'rather discourage women in his laboratory for research purposes'[21] and the environment at the Cavendish Laboratory could be similarly antagonistic to female researchers, exacerbated by overcrowding. J.J. Thomson, an assistant and then Professor of Experimental Physics there, recalled an episode in the 1890s that 'appealed to his sense of humour'. A 'lady student' from Newnham or Girton had fainted and a laboratory boy, 'anxious to rise to the occasion, thought it right to turn the fire hose on her!'[22] Women were in a precarious position whether they sought admission to research or teaching facilities, their admission was not by right and, often tolerated rather than welcomed, they were subject to the whim of male dons. For example, in 1887 Adam Sedgwick threatened to turn women out of his Cambridge morphology laboratory if they did not stop agitating for degrees.

Femininity in the laboratory: An uneasy combination

If laboratories held significance for masculinity, women's presence could be transgressive; in contemporary accounts this unease leads to a denial of women in the laboratory or their portrayal as objects of amusement. At Cambridge, women were not admitted until comparatively late to the University Chemistry Laboratory (1909) and were barred from physiology practicals until 1914. Richmond notes how women were excluded from the Cambridge Natural Sciences Club and from informal departmental and formal college tutorial systems and prizes. Ever enterprising, the women of the Balfour set up their own intellectual and social networks. This creation of parallel facilities, echoing and hand-in-hand with the provision of women's colleges, was one way that women sought to create a more feminised scientific environment. The Cavendish Laboratory at Cambridge has been described as 'a community built upon masculine ritual and tradition' and this is an apt description of the social culture under the regime of both J.J. Thomson and his successor, the more progressive Lord Rayleigh.[23] Despite this, female researchers did manage a degree of integration, mainly by dint of familial relations. Eleanor Sidgwick (one of the Balfour family, wife of Professor Henry Sidgwick) and Rose Paget (daughter of Sir George Paget, Professor of Physics) were part of the social network at Cambridge and their cordial family relations with Thomson or Rayleigh may have extended into the laboratory.[24] Neither of these women had been students at Newnham or Girton. Rayleigh's account of Paget's and Thomson's courtship shows, however, how inconsequential the work of women was represented:

> In 1889 Miss Paget did some research work at the Cavendish Laboratory on soap films thrown into stationary vibration by sound, after the fashion of Sedley Taylor's phoneidoscope. J.J. found that she needed some help in these experiments ... J.J. would come in, and say 'I must go upstairs for a few minutes', which very easily expanded themselves, by 'Relativity' perhaps, into an hour. One day he came down looking highly delighted, and Miss Paget went out with a flush on her cheek, and did not continue any more experiments ... After six weeks' engagement they were married ...'[25]

A few years later, Mrs. J.J. Thomson's role in the laboratory was typified during the opening of new laboratory buildings by the task of receiving the guests while holding a bouquet of flowers, the gift of the research students. It is likely that relations between the sexes in the laboratory were, for the most part, polite and cordial, but women were absent from the masculine image that the Cavendish and other research facilities wished to present to the public.

The above accounts of late-nineteenth and early-twentieth century laboratory life use models of moral courage and perseverance to construct a model of scientific manhood. It is difficult to position women within these narratives except as figures of amusement, oddity or romantic interest; it is easier to omit them all together. This is one reason why it is so difficult to get a clear picture of women's everyday experiences in the laboratory. Appreciations of Hertha differ from those of her male contemporaries by a lack of emphasis on her experimental work, despite her exhausting and sometimes dangerous research on large-flame electric arcs.[26] Even the announcement of the award of the Royal Society's Hughes Medal for original discovery in the physical sciences to Hertha in 1906 is notable for its brevity, the concentration on results rather than on any arduous process to attain them, and a lack of celebration – or even evaluation – of her work:

> Mrs Ayrton's investigations cover a wide area. She discovered the laws connecting the potential difference between the carbons of an

arc with the current and with the distance between them, and proved these to apply not only to her own experimental results but to all the published results of previous observers. Dealing with the modifications introduced into the arc by the use of cores in the carbons, she found the causes of these modifications. The peculiar distribution of potential through the arc was traced, and its laws were discovered by her.

Having found the conditions necessary for maintaining a steady arc, and for using the power supplied to it most efficiently, she was able to explain the cause of 'hissing', and the causes of certain anomalies in the lighting power of the arc.

For the past four years Mrs Ayrton has been engaged in investigating the causes of the formation of sand ripples on the seashore.[27]

Other individuals awarded medals at the same time as Hertha received praise for the value of their work and the method by which it was achieved. The recipient of the Copley Medal had provided 'distinguished services', 'brilliant memoirs' and was in the 'first rank of investigators'; the Royal Medal was awarded to Alfred Greenhill for his 'remarkable' work 'characterised by much originality, and by a rare power and skill' and to Dukinfield Henry Scott for his work which was of 'first-rate importance'. The recipient of the Darwin Medal was noted for 'extensive experiments that have been carried on for many years' with results 'of great importance' and 'considerable influence'.[28] As will be discussed in Chapter 7, the terseness of Hertha's announcement was possibly a reflection of discord within the Royal Society as a whole, and within the council, over the award of a medal to a woman.

Hertha's prize-winning work on the electric arc had been carried out at the Central Institute from 1893 to around 1904. It is difficult to follow her into the laboratory as her presence, arranged informally by her husband, is rarely noted in any existing archive of the institution. This relative 'invisibility' in college history is significant when it is remembered that Hertha's work here was aided by Central staff and students and led to the award of a prestigious Royal Society Medal – surely an important event for a college striving to win credibility as a professional institution on a par with the universities. Yet, as discussed in Chapter 3, it was important for the Central to prove itself capable of producing 'trained men' and 'professional engineers' who were able to maintain 'the whole status of the profession'.[29] This required promoting an image of itself as producing graduates with the robust (masculine) practical and leadership skills to build dams, construct electric light schemes or create new locomotive systems throughout the world.

This ideal of manly scientific citizenship was essential to the Central's efforts to establish itself as a provider of education that was equal (albeit different) to that provided by the long-established universities. A woman in the laboratory, no matter how successful her work, would not aid the promotion of this image.

A nervousness of showing women in the laboratory was not confined to the Central Institute. For example, an article on Girton College in a contemporary periodical was not atypical in showing well-equipped, tidy but empty laboratories.[30] The new women's colleges were wary of provoking criticism from opponents who argued that women jeopardised their well-being by following a programme of education similar to men's. Darwinian fears that exposing women to hard intellectual work could lead to a loss of health, fertility and beauty, were exacerbated by imagery that presented the laboratory as a harsh and competitive environment that threatened accepted norms of femininity. At University College London in the late 1880s, one female student was discouraged from seeking admission to chemistry classes as it was believed that women would be 'scarred for life and have their clothes burnt off them as the men threw chemicals around'.[31] The dangers inherent in the laboratory meant that it was an inappropriate place for a woman.

Hertha in the laboratory

Yet a glimpse of Hertha's experiences in the laboratory can be gleaned from the 1903 book which resulted from her researches.[32] Far from being a figure of amusement, it is clear that she established warm relationships with the staff and students who assisted her in experimental trials of carbons while following their diploma course. C.E. Greenfield, a student at the Central 1899–1903, later worked with Hertha in her home laboratory after her husband's death. Another co-worker at the Central, her husband's Chief Assistant Thomas Mather (who had worked with Ayrton on arc lights) rushed to Hertha's defence in a letter to *Nature* in response to an unsympathetic obituary which had accused her of a lack of originality.[33] In the preface to *The Electric Arc* Hertha thanks Mr. Mather for his valuable advice and assistance with experiments. Mather had superintended the first set of experiments and made suggestions which simplified their execution. In order to adequately monitor the carbons, a mirror was now used to reflect the phenomenon onto a large screen of white cartridge paper. This improved visibility was vital to Hertha's research which depended on observing the carbon and spotting minute changes over long periods of time. It was painstaking work which required keeping an arc under a steady current controlled by hand for periods of an hour or more while monitoring the results. (Hertha likened this process to the driving of an obstinate animal after learning its caprices.) Another of Ayrton's assistants at the Central, Mr. Phillips, made a similarly important contribution to the design of Hertha's experiments, introducing the use of a tubular carbon to ensure that the arc did not extinguish when air was blown on to it.[34] Effective teamwork and experimental facilities able to host lengthy, controlled experiments were indispensable. The relationships that Hertha achieved with co-workers at the Central were therefore vital to her work.

Challenging the pure and practical hierarchy

Given recent research revealing women in the laboratory, it is important to ask why their presence has for so long been obscured, not least by 'manly' representations of laboratory life of the kind described above. Despite the growth in number and prestige of laboratories (and the increasing recognition of their key role in achieving trade dominance) any hint of commercialism could still taint a researcher's work, as outlined in Chapter 3. In 1903 astronomer William Huggins, in his presidential address to the Royal Society, made implicit criticism of fellows (possibly aimed at Ayrton, his colleague Perry and their fellow modernisers) who worked not out of ideals of service but for profit, when he stated that the Society had achieved its unique position '... by its unwearied pursuit of truth for truth's sake without fee or reward'.[35] Men receiving money as a result of their scientific researches could jeopardise their credibility; for women to do so would be to transgress middle-class conventions of femininity and jeopardise social standing too. Emphasising the heroic qualities involved in research (perseverance, bravery, 'iron nerve') helped win prestige and moral currency at a time when science as a profession was still in the process of negotiating an identity and a space. For similar reasons, practical scientists and engineers constructed arguments for their own integrity based on their profession as providing 'good' to society. Modern engineers presented themselves as the bringers of progress, taming and domesticating physical forces and presenting them as a service to man (and woman) kind. Such discourse tended to imbue laboratory methodology – observation, experimentation, reason – with moral value, a view neatly presented in a Royal Society obituary which remarked that 'Scientific investigation is eminently truthful. The investigator may be wrong, but it does not follow (that) he is other than truth-loving'.[36] Here it is the method of science that bestows moral righteousness, not the correctness of any results that the experimenter may produce.

The merit of experimental work was also promoted for the valuable moral training it conferred. At Oxbridge and public schools, success in the laboratory or site of scientific experimentation was becoming as important as the playing field for the development of moral character;[37] and the adoption of an active, scientific methodology began to be presented as a paradigm of how the modern citizen should manage life. Everyday problems were to be approached with objectivity and clarity of thought – just as researchers in the laboratory approached their experimental tasks. If the laboratory was a space where manliness was developed and displayed, women's presence was problematic. For Conan Doyle's Sherlock Holmes (who was the epitome of precise scientific rationality) women were 'never to be trusted – not the best of them'.[38] This remark reflected views still articulated by some (but by no means all) followers of the new evolutionary understandings of man. Experimental research required the objective production and presentation of data, the capacity to abstract from the particular to the general, the intelligence to design, interpret and replicate experiments, plus the ability to focus without distraction. Darwinists such as Armstrong and Galton doubted whether women, who were designed by evolution to be emotional and partial, with a tendency to shift attention and achieve only surface understandings, could be trusted to do science. Women were competent in their 'natural' role as wives and mothers, but they were intellectually and temperamentally unsuited to scientific work: women's procedures and results simply could not be relied upon without question.

A connecting factor which helped to keep women at the periphery of laboratories was the practical sciences' competition with mathematics for status. The natural sciences tripos at Cambridge had been established in 1851 and, until the mid-1870s, it had been viewed largely as the poor relation to mathematics. In his memoir of J.J. Thomson, Lord Rayleigh (son of the Nobel prize-winning physicist of the same name) remarks in connection with the Cavendish Laboratory that prior to around 1865 the tradition of physical science at Cambridge had been somewhat antagonistic to experimental research. Since the 1870s however, all that had changed.[39] This process had continued and, by the mid-1890s, mathematics (which had long been a signifier of manliness) had lost its pre-eminent appeal for a majority of male students at Cambridge who, in preference, sat the natural sciences tripos. Furthermore, as the new women's colleges retained their preference for mathematics, with increasing numbers of students from Girton and Newnham taking honours in the mathematics tripos, the discipline became less antagonistic to contemporary notions of femininity. (This process will be explored in the following chapter.)

By the late nineteenth century, studying mathematics alone did not provide the endorsement of masculinity that it had earlier offered; nor did it promote the active masculinity now exhibited by the practical mathematicians, experimentalists and engineers in the vanguard of so much technical progress. Although the Cambridge mathematics tripos was still an important route of entry into experimental science until the late 1880s, the ending of the tradition of 'mixed maths', the infiltration of analysis, and the addition of higher mathematics to the natural sciences tripos, had all reinforced the perception of mathematics as less relevant to the physical world. This attitude is reflected in Lord Rayleigh's remark in a Royal Society Presidential Address that in some branches of Pure Mathematics it is said that readers are scarcer than writers.[40] Rayleigh's comment also conveys some of the unease and suspicion that can sometimes be engendered by a minority group engaged in producing inaccessible – even 'cabbalistic' – learning.[41] Engineers and experimentalists constructed an active, virile identity in the laboratory in part to highlight their difference from the pure mathematicians and so enhance their status.

Home laboratories and the gendering of space

Whatever the cordial relationships that individual women may have established with co-workers in the laboratory, their presence did not rest easily within the culture and image that male workers liked to present. Hertha undertook research at the Central laboratory only while her husband was professor there (her work a continuation of his experiments on the electric arc). After Ayrton's death in 1908, and the subsequent loss of his patronage and support, Hertha's tenuous and informal connection with the College was severed. As a woman, it would have been difficult for her to step into her husband's shoes or assume another teaching or research role at the College. When Pierre Curie died in 1906, even Marie Curie (who in 1903 was joint recipient of a Nobel prize with her husband and Henri Becquerel)[42] found it difficult to fulfil her hopes of continuing her work by taking over his professorship at the Sorbonne. Marie only succeeded in securing the appointment after special intervention by friends and supporters.[43] After Ayrton's death, Hertha transferred her attic laboratory into the drawing room of her London home and it was here that all her experimental work was undertaken until her death in 1923, fifteen years later.

Historians have written extensively about a late-nineteenth-century ideology that divided space according to gender and function. Separate spheres placed the (middle-class) woman in the home and men in the great world beyond; scholars of space have emphasised how women, as signifiers of moral virtue and family status, could not cross social and physical boundaries as men were able to do.[44] One of the strategies of the suffrage movement was to challenge this gendered occupation of space as women marched through the streets and took to public platforms to argue their case. As a member of the militant Women's Social and Political Union (WSPU), Hertha was closely involved with 'the Cause'; her campaign to be an equal member of the scientific community alongside her male peers ran parallel to her fight for the vote. Indeed, she saw both issues as separate facets of the same problem: the need for equality between the sexes. The controversial campaign of the WSPU, and the highly-public role assigned to Hertha within it, may well have added to the difficulties that she faced in gaining access to scientific institutions. It also added to reservations about the presence of women in the laboratory and made the issue even more contentious. As will be discussed in Chapter 7, the actions of the WSPU were regarded by many (even within the wider suffrage movement) as illogical, if not 'hysterical' – attributes which were coded feminine and seen as the direct opposite of the empirical, objective and unemotional methodology of experimental science.

Although the explanatory power of separate spheres has been increasingly challenged as providing too rigid a division, especially when applied to turn-of-century, it is clear that men and women shared few spaces on equal terms; women were defined by the home in a way that men were not. Although many male scientists had home laboratories, after the 'aristocratic house' tradition, by the late nineteenth century rarely were they the only venue for experimental work that these men had access to. That gentlemen's 'country house' laboratories continued at turn-of-century has been interpreted as a means to reconcile the idealism and spiritual value represented by the older scientific 'devotee' tradition with modern laboratories producing commercial applications alongside new knowledge.[45] This strategy, alongside that of representing the laboratory as a place of manliness and heroism, worked to assuage the new professional scientist's need to convey integrity and moral worth. Lord Rayleigh had developed large laboratories at Terling and a smaller one at his previous home in Cambridge, yet he also researched and held professorships at the Cavendish Laboratory and the Royal Institution, and was later associated with the new National Physical Laboratory. William Crookes experimented at home as well as undertaking joint work at the Royal Institution. George Gabriel Stokes carried out most of his research at a modest home laboratory but simultaneously held a Cambridge Professorship. It was the connection of these laboratories to a wider institutional and scientific network that helped preserve their credibility and integrity as viable places of experiment.

But these home laboratories were not all modest. Upon inheriting Terling Place in 1873, Lord Rayleigh installed a private gas works to feed bunsen burners and blowtorches, and developed the existing laboratory into a complex of experimental facilities. These consisted of a 'book-room', a workshop, a chemical room and a photographic darkroom. Above these on the main floor was the main laboratory, partitioned into a number of areas, including the 'black room' which was equipped with a helioscope and a spectroscope and used for optical experiments.[46] Crookes' private laboratories comprised a suite of rooms including a physical laboratory, chemical laboratory, workshop and library.[47] Experimental activities at these home laboratories were endorsed by the Royal Society of London of which all these men were fellows (and presidents in the case of Rayleigh and Crookes), a validation and honour denied to women. Working only from home, without position, implied amateur status by the closing years of the nineteenth century. After all, anybody could perform simple experiments in their home, as was encouraged by a series of articles entitled 'Science for the Unscientific' appearing in a popular periodical from 1894. Through experiments to be executed at home the series explored subjects such as electricity, air pressure and optical illusions.[48] Hertha's home laboratory contributed to her remoteness from the networks and standards of the scientific community and had similar effects to her exclusion from the Royal Society.

It seems likely that Hertha's laboratory lacked credibility as a viable experimental site at a time when increasing emphasis was being placed on precise measurement and the use of manufactured instrumentation.[49] The standardisation of values in areas such as electromagnetism and electrical resistance were of immense importance to securing the integrity of the international cable system and, as a result, extra demands were placed on the experimental space. In order to achieve exact results, laboratories were deemed to require specially-designed buildings and equipment which would aid replication of experiments and prevent any disturbance or contamination by the outside world which could distort findings. Arguments about the status of experimental findings could hinge on securing agreement on the acceptability of particular instrumentation, or on the viability of specific research technologies.[50] Whether experimental results were to be admitted as knowledge was now dependent on the credibility of their site of production, as well as on the trustworthiness of the 'men' of science who had produced them. By attempting to create a 'value-free' experimental space and quantifying and measuring their research findings, researchers attempted to give material expression to the 'objectivity' which was the lodestar of their experimental science. Just as these engineers, practical scientists and physical researchers were trying to dispense with the individual and present their conclusions as the 'view from nowhere',[51] their counterparts in pure mathematics, that most deductive of disciplines, were foregrounding the individual intellect and doing precisely the opposite. By promoting 'objectivity' in this way laboratories were striving for status as privileged epistemological spaces and replacing older venues of scientific enquiry such as the museum, fieldwork or the personal study.[52] When Karl Pearson took over Francis Galton's Eugenics Record Office at University College London in 1906, one of his first acts was to rename it 'The Galton Eugenics Laboratory', although much of its work centred on statistical computation, measurement and the compilation of tables for family pedigrees and actuarial death rates.[53] New institutional laboratories vied with each other for prestige based on their laboratory equipment. The Cavendish Laboratory had a 'magnetic room' for experiments in which there would be no magnetic disturbances from iron fittings or pipes. A 100-ton testing machine was pride of place at the new National Physical Laboratory, while the Central College's new electrical laboratories boasted a 5kw Ferranti transformer and a travelling overhead crane to assist in experimentation.

A woman's laboratory: Tensions and questions

It is significant that the quality and efficacy of Hertha's experimental apparatus was the focus of doubt and concern for the Royal Society on several occasions, leading to the rejection of a paper on sand ripples.[54] One referee questioned her research methods and characterised them as 'crude'.[55] In one of her papers on the subject, Hertha describes her methods. Initial experiments were executed with vessels of various shapes and sizes from a 'soap dish' to a 'pie dish', moving up the scale to a 'tank' forty-four inches in length. Rollers or cushions were put under these vessels to enable smooth rocking (by hand or small electric motor) which caused the water oscillations. In this way she examined varying ripple formations and, by adding finely-powdered aluminium to the oscillating water, made visible the characteristics of water vortices.[56] Hertha's experiments demonstrated that ripple marks were not formed by friction (as put forward by Professor George Darwin at Cambridge University) but were due to the processes of varying water pressure. According to Tattersall and McMurran, in the 1940s independent experiments with ripple tanks by Bagnold and Scott confirmed Hertha's conclusions, yet neither mentioned her pioneering results.[57] If Royal Society referees questioned her techniques, even sympathetic contemporaries expressed mild scepticism and wrote of how hard it was to appreciate her ideas due to the 'toy-like models' used in her laboratory.[58] Another memoir describes Hertha's use of 'a morsel of feather on a single thread of silk, anchored to a hat-pin' with which to test the speed that coal gas is driven through tubes.[59]

Yet, there is a gender difference here. Male scientists were applauded for their achievements gained with primitive apparatus, especially when these heightened the danger to the experimenter. George Gabriel Stokes was celebrated for experimental work 'executed with the most modest appliances' and Rayleigh's Terling laboratory was famously said by Kelvin to be 'held together with sealing wax and string'.[60] John Aitken, working at his home laboratory in Falkirk, was praised for 'researches devised and constructed in his own hand'; his work tables were 'laden with glass-work, blow-pipes and many odds and ends of apparatus in the course of construction' yet, far from employing 'crude' methods, he possessed 'a real intuition for the devising of illustrative experiments. His one aim was to get at the truth'.[61] Such characterisation referred back to the romance of the earliest experimenters, such as those who founded the Royal Society in the 1640s, and linked the contemporary 'great men' of science with their forefathers. The tendency to highlight these older, more 'hands-on' experimental techniques may also have been a way for modern experimenters to assert their heroic credentials in the face of the increasing role of instrumentation which functioned as a barrier, and an intermediary, between 'men' and their science.[62]

Despite attempts to depersonalise and objectify the experimental space, trust was intimately connected with the researcher whose space it was – and that researcher's body was gendered. Gender has been notable by its absence from the recent debate on 'trust' in the history of science (an omission also discussed in Chapter 7) but it is clear that women, mostly excluded from institutional and personal networks, and facing a Darwinian question mark over their intellectual capacity for science, did not easily acquire the credentials for 'trustworthiness'. Shapin and Schaffer, in the context of the seventeenth century, have argued that the process by which a piece of work becomes accepted as knowledge rests on relations and conventions of trust between disinterested (and like) gentleman scientists.[63] Another view interprets the increasing importance of objective and transparent science, based on measurement and precise mathematics, as a way to replace any reliance on the skill, good intentions and veracity of the practitioner.[64] On both counts Hertha was on the losing side. Her laboratory did not have the credibility of a modern, 'value-free' experimental space; her sex made it difficult to forge working relations with male peers as she was barred from institutions such as the Royal Society which were vital for networking and reputation-building. Both of these difficulties informed each other, escalating the scepticism that Hertha faced as a female scientist and influencing the kind of investigations that she could perform.

Hertha was limited in the type of work that she could undertake in her laboratory which was simply not equipped to perform, say, electrical experiments requiring precise measurement. She found it increasingly difficult to persuade contemporaries to visit her home to witness her experiments and this was especially acute during the first months of World War One when Hertha struggled to have any official interest taken in her anti-gas fan. One commentator, a President of the Institution of Electrical Engineers, refers with ambivalence to Hertha's laboratory as a 'laboratory-drawing room'.[65] A photograph of 'Hertha Ayrton in her laboratory' which, although undated, probably originates from 1906, the year that she received a Royal Society Medal, endorses this perception (Figure 5.2). Hertha is positioned in front of a bookcase, a potted plant and vase are above each shoulder and paintings hang on the wall above her head. She stands in front of a table upon which is a barely visible glass tank. The edge of another glass tank can just be seen, resting on top of a table covered in a velvet cloth. Hertha herself is dressed as to receive visitors, wearing jewellery, avoiding our eyes by gazing out towards the right of the photograph. The apparent domesticity of her experimental apparatus means that there is no signifier of Hertha's profession in the portrait. The effect is ambiguous: is this a scientist in the laboratory? Or a hostess/woman in her drawing room? The tidy, domestic values so connected with notions of femininity can also be read from this image. Compare this to the messy, busy, active photographs of her

Figure 5.2   Portrait of Hertha Ayrton in her laboratory (1906) © The Royal Society

contemporaries' laboratories (Figures 5.1 and 5.3) and the visual subtext revealed by Hertha's portrait is that a woman's space is the home, not the laboratory.

Figure 5.3 William Crookes in his laboratory. Images from Lord Rayleigh, Robert John Strutt, 'Some reminiscences of scientific workers of the past generation, and their surroundings', *Proceedings of the Physical Society*, 48 (2) (1936)

When Hertha Ayrton's achievements were reported in the press, the emphasis was on her femininity as a source of wonder and amusement. For example, *The Times*, reporting on the award of the Hughes Medal to Mrs. Ayrton, noted that her experimental methods were to be appreciated 'with interest and entertainment'.[66] In her biography of her friend, Evelyn Sharp cannot resist using Hertha's experimental enthusiasms as a means to introduce levity into her account. On the occasion of a dance at Hertha's home, hosted by her stepdaughter for her friends, Sharp writes that the

> guests were suddenly amazed to see an abstracted lady in an overall emerge from the laboratory at the top of the house (where she was using a good deal of current at the moment) and, muttering to herself, 'The main fuse will go!' proceed to switch off the electric lights without apparently noticing the presence of the revellers.[67]

Even a sympathetic novel, written by Edith Ayrton Zangwill and inspired directly by the life of her stepmother, uses the spectacle of a woman in the laboratory as a curiosity with which to generate humour:

> From the room within came a curious fizzling sound and a faint but still more curious odour. Some demented domestic appeared to be frying a late and unsavoury lunch in her bedroom. No servant would have condescended to a shapeless, blue-cotton overall and, still less, to hideous, dark goggles, made disfiguring by side-flaps ... all was dominated for the moment by a hissing jet of flame that darted out between the small, dark objects held in metal clamps which stood on a table in front of the girl ...[68]

The ambivalence of a woman in the laboratory is a theme of the novel, represented by the heroine's mother who names her daughter's laboratory the 'infernal regions' of the house and feels uncomfortable visiting there. The heroine's suitor also finds accompanying Ursula/Hertha into the laboratory personally uneasy. He remarks that he prefers to meet her in the park because she is 'less scientific, more human, more personal' – by implication, more 'womanly' out of the laboratory than within it.[69]

Zangwill was not alone in presenting women and the laboratory as an unsettling combination; H.G. Wells has the suitor of his heroine, Ann Veronica, call her experimental equipment and materials 'All your dreadful scientific things'.[70] *The Call* was published in 1924, the year following Hertha's death. The scientific experiences of Ursula in the novel follow closely those of Hertha (although the former's family situation is very different to conform to the requirements of a romantic novel). Passages of Evelyn Sharp's biography echo word for word episodes in *The Call* and Sharp and Zangwill were in correspondence with each other. Sharp's *Memoir* also betrays some ambivalence to Hertha's laboratory, this time from her natural daughter, Edie, who is quoted as complaining 'I do wish mother had a boudoir all filled with yellow satin furniture, instead of a laboratory ... like the mothers of other girls'.[71] In the novel Ursula is a Christian (nominally at least, as the fictional family celebrate Christmas) so Hertha's Jewishness is ignored, despite the fact that the author was married to leading Jewish writer Israel Zangwill (who was, therefore, Hertha's son-in-law). Was this a deliberate attempt to downplay Hertha's Jewish heritage? Or merely Zangwill following the conventions of romantic fiction? (And if the latter, why did Zangwill feel that having a Jewish heroine would contravene these conventions?) It is difficult to assess whether Hertha's ethnicity, once she had downplayed Judaism on her marriage, played any part in her later marginalisation from male scientific networks. Certainly, for Tattersall and McMurran, Hertha's 'Jewish background does not seem to have affected her social or scientific life',[72] yet it is difficult to determine the inner motivations of those who chose not to include her. It will be suggested in Chapter 7 that Hertha's prominent Jewish features, especially in her later years, may have led to her identification with the threatening image of the 'black widow', a representation used to question women's rationality, especially that of older 'militant' women fighting for female suffrage (a cause with which Hertha was associated publicly).

Women's place in science

As the institutionalisation and professionalisation of science consolidated in the first two decades of the twentieth century, a place in science, and by extension in the laboratory, was negotiated for the increasing number of women with scientific qualifications. Scholars including Margaret Rossiter and others have demonstrated how women were incorporated into scientific employment and research but were segregated within it to roles that made use of their special 'feminine skills'. These roles were typically low-paid, low-level technician and assistant appointments which worked to reinforce the dominant gender hierarchies in science, not to challenge them. They often required 'great docility or painstaking attention to detail' and writers calling for increased opportunities for women 'glorified these positions, considered them very suitable for women in science, and advocated more of them'.[73] Rossiter locates most of her research in the United States, but similar processes occurred in the UK, accelerating during World War One when women were needed to replace men away on active service. It was during these years that women first gained admission to the National Physical Laboratory where they were employed in the Metrology Department to undertake gauge-testing, an important aspect of the munitions industry. By the summer of 1917 there were ninety-nine women out of 420 staff in the department.[74] As considerably more female scientific graduates entered the British labour market, so laboratory jobs requiring patience, persistence and a capacity for tedium became gendered 'female'. Such a distinction is not easily identifiable during the nineteenth century however as at that time men undertook – and were highly praised for – this kind of work.

Precise measurement was becoming increasingly important to science, an approach led from the mid-nineteenth century by William Thomson (Lord Kelvin) at the University of Glasgow[75] and adopted later by Cambridge physicists at the Cavendish under the leadership of Lord Rayleigh. Isobel Falconer has demonstrated that nearly all the experiments undertaken at the Cavendish were 'finicking and difficult to perform' calling for patience and tenaciousness from all researchers.[76] Certainly this was true for Lord Rayleigh and Eleanor Sidgwick's experiments to derive standards for the Ohm. These were conducted late at night to avoid magnetic and other disturbances, Rayleigh regulated the speed, Dr. Schuster took the main readings and Sidgwick recorded the readings of the auxiliary magnetometer. Equipment consisted of an electrically-driven tuning fork and a circularly-calibrated card attached to a spinning coil to confirm that the rotational speed coincided with the frequency of oscillation of the tuning fork. A critical feature of the experiment was the speed control for the rotation of the coil; a secondary needle was introduced to monitor the effects of changes in the direction of the Earth's magnetic field.[77] When Dr. Schuster left Cambridge, his part of the work passed to Sidgwick, while hers went to various volunteers, including her sister and once to her brother, Arthur Balfour. The division of duties here was not made along gender grounds. Important criteria for assuming responsibility for taking measurements was possession of scientific knowledge and aptitude for teamwork, based in this case on familial relations.

Moving away from the Cavendish, Peter Broks has written that 'patience' and 'perseverance' were bywords for scientific practice in the late-Victorian and Edwardian era, used to describe scientists of the day to such an extent that they 'verged on cliché'.[78] Such frames of reference were not used solely by journalists; Rayleigh praised John Kerr for his discoveries in electro-optics 'accomplished under no small difficulties by courage and perseverance' and the journal *Nature* praised William Crookes for the 'extreme care and pains' that he took in making weighings'.[79] What is significant here is that around 1900, the signification and meaning of laborious and painstaking work began to change according to the sex of the researcher who performed it. By the end of World War One, routine, 'finicky' work was largely the preserve of women.

At the turn of the nineteenth century, there were few female researchers and the whole practice of science, and its significations, were still in the process of being negotiated. In this fluid and changing situation, women did not yet have a fixed, prescribed role in the laboratory. The experimental facility was presented as a masculine place with an abrasive culture, symbolic of the manly virtues of bravery, persistence and fortitude, and femininity was not included in this image. However, this is not to deny that some of the women who entered the laboratory extended existing ideas of women's service to the work that they performed there. William Crookes's wife was said to be delighted to help him when she could, often carrying out weighings for him in his chemical work which 'as she said, was suited to the delicate fingers of ladies'.[80] Crookes did most of his experimental work in his home laboratory, so his wife did not need to leave the domestic sphere and her work alongside her husband could be presented as a natural extension of her wifely role.

Conclusion

As historians of science and gender have illustrated, the accelerating professionalisation of science at the end of the nineteenth century did pose an effective barrier to women's participation. Hertha found obstacles throughout her life which made it difficult for her to sustain an identity as a professional scientist like her many male peers. However this process operated in a more complex way than just the closing of institutional doors to women. Justification and naturalisation of women's varying exclusion from the laboratory required the propagation of a masculine laboratory culture – both actual and mythic – which emphasised the physicality of the experimental space and the dangers that could lurk there. The creation of an heroic identity for scientific experimenters also served a useful function in elevating their moral and material worth at a time when the authority and status of their profession was being established. It is interesting to note that the experimental and physical sciences were, at this time, removed from nature to the laboratory while, in contrast, sciences such as botany and geology continued in addition to seek phenomena as they occurred in nature. Women have been associated with the natural biological sciences while becoming virtually invisible in the physical ones: that the laboratory was configured as an unsuitable place for a woman may have contributed to this gender divide.

6
The Mathematics of Gender: Women, Participation and the Mathematical Community

When Grace Chisholm Young published a joint paper with her husband in the *Proceedings of the London Mathematical Society* in 1910, she was just the fourth woman to have published there since the journal's creation in 1865. By this time female membership of the London Mathematical Society was around 5%, the majority of whom were women who had connections with the new colleges for women at Cambridge. Both Girton and Newnham had produced a number of notable female wranglers in the Cambridge mathematics tripos in the years surrounding 1900, plus a far greater number of women achieving second-class honours. When it comes to assessing the participation of women in mathematics, as these preliminary remarks suggest, whichever variable chosen will depict a differently-focused account. Should we look at the membership of learned societies? The number of mathematical publications by women? The number of women progressing to the specialist part II of the Cambridge mathematics tripos or gaining a higher degree in the subject? Or the number of female academics in mathematics perhaps? Each of these perspectives will suggest a subtly different picture.

This chapter will explore the nature and scope of women's involvement in mathematics in the decades surrounding 1900, putting that participation into context and investigating the confluence of conditions that facilitated a female contribution at a time generally held to be sceptical, if not hostile, to female rationality. It will demonstrate that women were accepted into the mathematical community – into academic networks and learned societies – with less hostility than their scientific sisters, and that this was made possible by key differences in practice and culture between the two disciplines. Mathematics did not professionalise in the same way as the practical or experimental sciences and retained a strong tradition of amateur involvement. This opened up opportunities for women that remained closed in scientific communities with a professionalising agenda and members' material, as well as scholarly, interests to promote. Moreover, in seeking its legitimacy from purity and distance from the material world, pure mathematics became less conflicted with contemporary norms of femininity; a process that developed, in part, in response to the masculinised culture of bravery and stoicism that emerged around experimental science in the closing decades of the nineteenth century. As material practices differed, so did each discipline's use of language and metaphor, and these new 'colourings' had the effect of naturalising women's inclusion or exclusion. However, despite the mathematical community welcoming women at one level, it excluded them from its highest honours which were achievable only by men. Fuelled by an emphasis on individual creativity, pure mathematics placed a romantic tradition of genius at centre stage; an ideology which privileged the male mind and ensured that gender-specific hierarchies remained in place even as women were accepted into the mathematical community.

Women and mathematical education

There is no doubt that the growth of higher education for women was a turning point in women's access to higher-level mathematics as the universities enjoyed a near monopoly on providing advanced mathematical education. By 1900 there were a number of institutions in Britain where women could study mathematics to degree-level (although at Cambridge and Oxford Universities women were allowed to sit examinations only, not to be awarded degrees). In addition to Girton and Newnham Colleges at Cambridge and Somerville and Lady Margaret Hall at Oxford, in London women were catered for by Bedford, Westfield and Royal Holloway Colleges, as discussed in Chapter 1. When the new civic universities were established they opened their doors to women from the start. The Universities of Manchester, Liverpool and Leeds emerged from the Federal Victoria University which had received its charter in 1880 and which had allowed women students from the beginning; similarly Birmingham (1900), Sheffield (1905) and Bristol (1909). The University of Durham was established in 1832 but did not welcome women until 1895. Were women studying mathematics at these new civic universities prior to World War One? Not at Leeds, where the first woman to graduate in mathematics did so in 1922, nor at Sheffield where the first female mathematician graduated in 1924. But these figures may point a too pessimistic picture if taken in isolation as they only include women taking *degrees* in mathematics, not mathematics *courses*. At Liverpool for example, where four women graduated in mathematics up to 1914, a mathematics module was a popular choice as a component of an arts or science degree. At the University of Manchester, sixteen women graduated in mathematics to 1914.

The four universities in Scotland (Aberdeen, Edinburgh, Glasgow and St. Andrews) admitted women from 1896 and, to 1914, they had produced collectively eighty-five women mathematics graduates, a few of whom joined the Edinburgh Mathematical Society. The University of Wales received its charter in 1893 and, to 1914, just seven women are recorded as graduating in mathematics. In Ireland, Trinity College Dublin agreed to admit women to degrees in 1904 and four women graduated in mathematics to 1914; for the National University of Ireland, which gave women access from 1878, the figure to 1914 is fourteen; at Queens College, Belfast, five women graduated in mathematics between 1908–1914.[1]

Despite these figures, research opportunities, so fundamental to women's ability to contribute to creative mathematics, were still very limited. For a man, success as a top wrangler in the Cambridge mathematics tripos may give him access to a college fellowship, eligibility for mathematical coaching work and a continued, intimate connection with the mathematical community of his university. He would also be eligible to compete for the Smith's Prize, a competition designed to encourage deeper mathematical thinking than that required for the tripos and seen as an important gateway to an academic career. These opportunities may not have been plentiful (and a high position in the tripos was also an entrée into elite professions beyond academia, a route that many senior wranglers took) but for top women the situation was immeasurably more difficult.[2] The women's colleges did not have the financial benefactors or resources to fund fellowships and lectureships in the same manner as the older, well-endowed Oxbridge colleges and, as a result, the growing number of women students entering British universities was not matched by a commensurate increase in the appointment of women teachers at those institutions.[3] Women's colleges too, were isolated from the scholarly and power networks that operated within the ancient universities. Female dons and 'graduates' were not awarded degrees or full voting/membership rights on the same terms as the men until 1920 (Oxford) and 1948 (Cambridge) and this meant that they had little say in university reforms and were dependent on the goodwill of male dons for providing teaching and, in particular, supervision for research.

As touched on in previous chapters, despite the barriers to graduate research for women, by the end of the nineteenth century a small but growing number of female mathematics wranglers had managed to achieve doctorates and university teaching positions. Girtonian Charlotte Angas Scott spent four years as a residential lecturer at Girton while studying for a DSc with Cambridge's leading pure mathematician, Arthur Cayley. Her doctorate (1885), as her official BSc (1882), was awarded by the University of London as Cambridge did not give degrees to women. Philippa Fawcett, who had famously been placed 'above the senior wrangler' in the 1890 tripos, was awarded a one-year research scholarship at Newnham after which she was appointed lecturer. In 1892 Fawcett was also made a fellow of University College London. Theodora Margaret Meyer lectured in mathematics at Girton from 1888–1918, becoming director of studies in mathematics in 1903; Frances Cave-Brown-Cave followed a similar trajectory, moving from success in the tripos to being a member of staff from 1903–1936. Meyer and Cave-Brown-Cave forged long-term teaching careers at Cambridge but did not make significant contributions to research. Similarly, Fawcett only ever published one research paper and retired after a few years to undertake education/ governmental administration; Scott, like so many of Cambridge's top-performing female mathematicians, found it necessary to venture abroad in search of opportunity. At Bryn Mawr College for women in the States she made a marked impact on women in mathematics by supervising several women's doctorates, including that of fellow Girtonian Isabel Maddison. Maddison remained all her life at Bryn Mawr as a lecturer and, increasingly, an administrator.

The small number of women finding opportunities in research and university teaching did not reflect the increasing number of women gaining a high place in the Cambridge mathematics tripos and going on to sit the specialist extension examination taken at the end of the fourth year of study. Women's names were first listed on the annual order of merit in 1882 but, by this time, the tripos was beginning to come under attack by reformers who believed that its highly competitive (and highly public) ranking system was detrimental to the development of mathematics. The next three decades were a time of controversy and change as Cambridge reformers sought to modernise the tripos in an attempt to encourage creative research which, it was argued, was being sacrificed to examination technique as candidates trained solely for a high place on the lists. These pressures resulted in an initial change in 1882 when the examination was divided into part I (four days of examinations) and part II (five days) which were taken a week apart in the summer term after three years of study; it was the results of this that comprised the annual order of merit. Progression to a new, specialist part III, open only to 'wranglers', was taken in the tenth term. In 1886 this was simplified to part I, taken over several days and resulting in placement on the order of merit, followed by the more specialist part II as a separate examination taken by 'wranglers' only. This latter examination enabled candidates to specialise in pure or applied subjects in their fourth year by choosing in advance the division in which they wished to be examined. These changes took place alongside complementary reforms in the natural sciences tripos which led to an increase in its mathematical component and it beginning to rival the mathematics tripos as a key route into laboratory-based science.[4]

The changing relationship between the mathematics and natural science triposes is significant in addressing women's participation in both. In 1883 a special provision allowed graduates from the mathematics tripos to proceed directly to the advanced part of the natural sciences tripos and, by end-of-century, so many were taking advantage of this that the decreasing numbers of men taking the specialist part II of the mathematics tripos became a serious concern to Cambridge's pure mathematicians. By 1900, the natural sciences tripos included higher mathematics and had become a strong examination in its own right. In contrast, a de-emphasis of physics in the mathematics tripos had made that examination less important for physicists. The implications of these developments for the mathematics tripos were made explicit by J.W.L. Glaisher, then president of the London Mathematical Society, when he announced that the 'senior wrangler is displaced from his throne' as he no longer owed his position to the results of the whole examination. As a result, the titles of wrangler, senior optime and junior optime had 'lost their old significance'.[5] Earlier, on the introduction of these reforms, the *Girton Review* had explained that in order to take part II, a candidate must first pass part I, adding that 'the standard being, however, so absurdly low that Mathematical Honours are notoriously the easiest road to a degree'.[6] The inclusion of women's names on the order of merit from 1882 is usually credited to pressure resulting from the success of Charlotte Angas Scott in being placed equal to eighth wrangler in the 1880 mathematics tripos. However, that women's inclusion had coincided with the introduction of changes that radically reduced the prestige of a top place in the lists made this concession one that was less threatening to grant and, at the same time, turned it into a somewhat pyrrhic victory.

Mistresses of theory? Women on the mathematical pass lists

Women's names did occasionally feature among the top wranglers: to 1910 there were eight women from Girton and Newnham placed in the top ten, and another three in the top fifteen.[7] These figures do not include the many women placed in the twenties, or Charlotte Angas Scott's achievement in winning a place equal to the eighth wrangler in 1880, as this was before women were listed formally alongside the men. Adding women into the equation extends any understanding of the dynamics behind changes in the tripos and poses questions for arguments connecting mathematics, sexuality and manliness in the last decades of the nineteenth century. The female students may have been peripheral in terms of numbers, but their impact on the prestige, culture and configuration of the tripos was substantial – not least *because* at mid-century the examination was (as Warwick demonstrates[8]) perceived as so essentially masculine. That women were achieving high positions on the pass lists was one factor in lowering the prestige and status of the examination, a process that accelerated from the 1880s to the abolition of the order of merit in 1909.

From 1882 the successes of the women's colleges were highly visible, both within Cambridge and beyond, as women's names were now being announced alongside the men's at Senate House and being reported in local and national newspapers such as *The Times*. Reformers of the tripos system centred their arguments on claims that the order of merit encouraged technique at the expense of real mathematics and, to bolster their case, they pointed to women's increasingly good performance as evidence of the examination's shortcomings as an indicator of real, creative mathematical ability. As outlined in Chapter 1, an important element of contemporary understandings of the female intellect was that women were intuitive and quick in gaining a shallow understanding of a subject, but were inferior to men in their power of reasoning, creativity and stamina for study. Women were diligent 'followers' but their lack of originality made it impossible for them to do more than travel in men's footsteps. According to this view, women's success in the tripos could only mean that the examination itself had been 'dumb-downed'. Women's success was one factor in pushing the mathematics tripos off its pedestal as the elite degree for men; another was the introduction of part II which marked the beginning of the end of Cambridge 'mixed mathematics' and created the conditions for the growth of pure mathematics as a distinct, separate subject.

So what was the performance of women in the Cambridge mathematics tripos as a whole? Analysing the data from 1882 when women were first included, to 1909 when the order of merit was abolished, reveals that there were thirty-three female wranglers and 385 honours passes at senior optime and junior optime levels from Girton and

Newnham. The year that Philippa Fawcett beat the senior wrangler was also notable for producing two Girton wranglers (both placed in the twenties) whose success has been inevitably overshadowed. These numbers are not an insubstantial achievement as women suffered the handicap of inadequate preparation and only variable access to top-flight coaching. (See the discussion on coaching in Chapter 1). Significantly, by the 1890s women were keen to sit the more specialist part II of the mathematics tripos. This examination was open – usually, there could be special dispensations – only to wranglers, therefore the number of women participating was necessarily small; when Grace sat the examination in 1892 she was the only woman alongside fourteen men. With the growth in popularity of the natural sciences tripos however, the numbers of men entering for part II also diminished and women comprised over a third of all candidates in some years. For example, in 1899 three women and four men passed the part II examination; in 1901 the figures were five and two respectively. In the years that lists are available, twenty-four women achieved a pass in this advanced extension examination. Given the diminishing entry of men, a higher proportion of women wranglers proceeded to part II than did their male counterparts.

Mathematics or natural science: A gendered choice?

A gender comparison of candidates achieving part 1 of the mathematics tripos and the natural sciences tripos clearly shows the latter becoming dominantly masculine in character as women comprise a growing proportion of candidates in mathematics (see the graphs in Figure 6.1). The trend by male wranglers to move to mathematical physics and the natural sciences can also be seen in essays submitted for the annual Smith's Prize (although this evidence is indicative only as just small numbers of each year's most-distinguished wranglers competed). The Smith's Prize was established in 1768 by the will of Robert Smith who provided for two annual prizes for proficiency in mathematics and natural science. Originally awarded via an examination, in 1883 the rules changed to allow competitors to submit an essay on a relevant topic of their choice; a change which was made in part to give candidates better opportunity to show deeper mathematical skills. This development added to the reputation of the Smith's Prize for providing a far better test of a candidate's capacity for insightful, creative mathematics than the mechanical, problem-based tripos. Indeed the encouragement the Smith's Prize gave for deeper mathematical thinking has

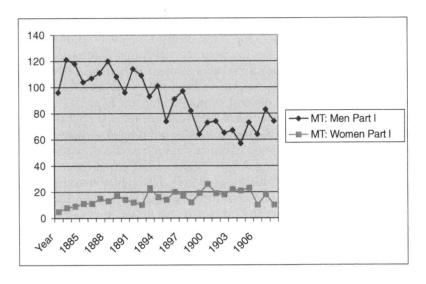

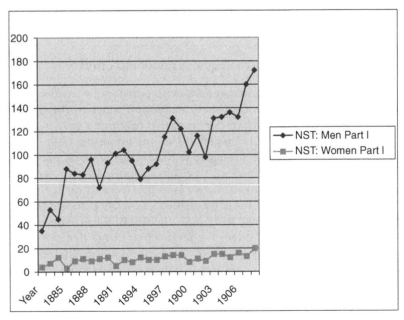

Figure 6.1   Number of candidates for the mathematics and natural sciences triposes at Cambridge, 1880–1910

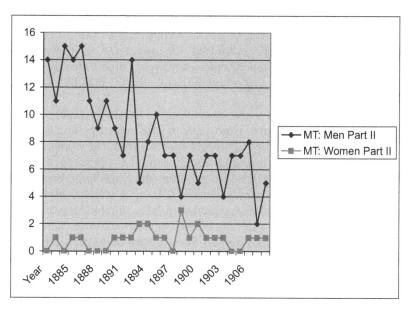

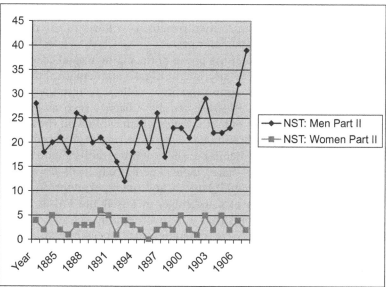

Figure 6.1 Number of candidates for the mathematics and natural sciences triposes at Cambridge, 1880–1910 – *continued*

been credited as partly responsible for the great flowering of mathematical physics at Cambridge in the second half of the nineteenth century.[9] Of the essays submitted for this Prize in the period 1885–1900, pure and applied topics were evenly distributed; between 1901–1920 however applied essays outnumbered the pure by a ratio of approximately 3:2.[10]

Figure 6.1 also shows women becoming increasingly visible in the Cambridge natural sciences tripos, due in part to the opening in 1884 of new laboratory facilities to prepare students from the women's colleges for the examination, and also due to the effects of William Bateson's centre for research into genetics which was largely 'staffed' by women from Newnham.[11] However, natural science, especially physics, was not universally accepted as quite as suitable for women as mathematics; in the first decade of the twentieth century there were suggestions at Newnham to replace science with domestic science – a development which was dismissed angrily by researchers such as Ida Freund.[12] Similar moves were also made – this time successfully – at the 'Ladies' Department' of King's College, London which, in 1915, became the 'Department of Household and Domestic Science'.[13]

Despite the increased visibility of women in natural science at Cambridge, the ratio of women to men remained far lower in the natural sciences than in the mathematics tripos.

It was at this time too that the power of the mathematical coach was being curtailed as the university sought to take mathematical instruction 'in-house' with the appointment of college lecturers and the growth of university-wide lecture series.[14] Coaches such as Edward Routh had been the focus of mythologies which centred on the driven competitiveness of their training and the male rivalry of the crowded coaching room. The slow demise of the coaching tradition and subsequent diminution of this shared male experience helped, along with women's infiltration into mathematics and the rapid adoption of the natural sciences by men at Cambridge, to lessen the link between mathematics and manliness that was so strong mid-century. Even if actual numbers of women were sometimes small (specifically in part II) their larger presence proportionally in the mathematics tripos helped to inform a new impression of the examination as one that was more at ease with 'the feminine' than natural science; an impression which added to doubts about its quality and prestige.

At the University of London however, particularly at Royal Holloway and UCL, women were taking up the natural sciences with enthusiasm and success. Marie Stopes (1880–1958) was just one of the women who gained a London BSc and went on to contribute to science teaching and research; she gained her degree, with honours in botany and geology, in just two years and pursued a successful career in paleobotany before becoming submerged in the birth control movement. Many of the women listed in the succeeding chapter as contributors to the *Proceedings of the Royal Society* held degrees in the natural sciences from the University of London. Despite this, there is some evidence to suggest that women students favoured pure mathematics in relatively greater numbers than their male peers. Elizabeth Larby Williams, who entered the University of London in 1911, remembers that pure mathematics courses were not in demand and in order to provide a sufficiently large audience there was a combined programme of lectures for students from Bedford, King's and Westfield Colleges. Even so the class was barely twenty strong and there were almost as many women as men. Women's choice to pursue pure mathematics over applied may not have been an altogether free one however. Larby goes on to reflect on women's lack of preparation for mathematical study, a lack which limited their options. Women were 'sorely handicapped by [their] lack of knowledge of applied mathematics'; none of the women had taken any public examination in applied mathematics and their knowledge of physics was very limited. Larby illustrated her point with her own experiences at East Ham Girls' Grammar School where she learned just heat but no mechanics, electricity, sound or light.[15] Evidence for some female bias towards the pure can also be discerned at Bedford College. In the academic year 1910–1911 for example, nineteen classes a week were scheduled in pure mathematics (algebra and trigonometry; geometry and calculus) and eleven in applied mathematics (statics; dynamics and hydrostatics), one of them a practical which ran fortnightly.[16]

Beyond university

What career choices were available to mathematical women seeking to make best use of their talent beyond university? Teaching was a conventional choice and many mathematics graduates subsequently taught in the burgeoning girls' high schools including, notably, Sara Burstall and Larby Williams. There was a small community of mathematics mistresses who shared strategies for teaching mathematics to girls and, in December 1911, a number of them met at a meeting of the London Branch of the Mathematical Association to debate the issues. A paper by Burstall (then Headmistress of Manchester High School for

Girls) was read and this, plus ensuing discussions, were published in the *Mathematical Gazette* the following January.[17]

Another, more limited, option for women mathematicians was to undertake complex calculations and statistical analysis as a 'computer' in a research laboratory, observatory or business such as insurance. In astronomical observatories in England, these jobs were generally taken by young boys, fresh out of school, who received miserly pay to undertake tedious, routine mathematical calculations. In the early 1890s however, a special scheme was initiated by Sir William Christie, Astronomer Royal, to recruit women mathematicians as supplementary 'computers' at the Royal Observatory in Greenwich. Unlike their male peers with equal mathematical skills, it was believed that these women may accept low-level posts which were both temporary and extremely low paid.[18] Two Girton graduates were amongst the first few women to take advantage of this short-lived scheme, Alice Everett and Annie Russell Maunder, who both took the Cambridge mathematics tripos in 1889.[19] Everett was offered a post at just £4 a month and was dismayed at this small sum:

> 'Does the fact that I have taken the mathematical tripos at Cambridge make no difference?' she asked. The reply was to the effect that she might take it or leave it. She had no alternative: she accepted the post and, with a stipend less than Caroline Herschel's a century earlier, began work on September 1, 1891.[20]

Despite these exceptions, this route did not immediately make it easy for women to become professional astronomers, especially in Britain, as has been documented by Peggy Aldrich Kidwell.[21] By tradition, British astronomers studied mathematics at Cambridge before gaining posts as assistants at observatories or working as directors in colonial observing stations. This 'learning on the job' was not easily accessible to women; it was acceptable for female students to study in the safe confines of a women's college but much less so to take a paid job in a traditionally male environment. Kidwell notes that the British practice of teaching astronomy as part of mathematics and physics (there was a telescope at Girton of which Grace made use during her time as a mathematics student there) served to limit rather than open opportunities for women. When the women's colleges hired teachers they chose mathematical women and not women astronomers, so women had little incentive to specialise. This was in marked contrast to the United States where important women's colleges developed departments of astronomy that trained women students and recruited female astronomers on to the teaching staff.

Another limited option for mathematical women was to apply their skills to statistics. In 1891, Karl Pearson had co-founded the science of biometrics which applied statistical analysis to eugenics and evolutionary research, following Francis Galton's methods. Pearson employed a number of women at various times in his Biometrics Laboratory which he set up at University College London in 1903. Among these women were Girton 'graduates' Frances and Beatrice Cave-Browne-Cave (who combined this work with teaching) and Julia Bell, who had studied mathematics at Girton but failed to sit the tripos due to ill-health. Another Girton mathematician, Florence Tebb Weldon, acted as a computer in collaboration with her husband, the biologist W.F. Weldon, who was involved with Galton's initial biometric statistical initiative.[22] Other notable researchers and biometric computers were Alice Lee who had studied mathematics at Bedford College and went on to gain a DSc under Pearson's guidance, and Ethel Elderton who came from a mathematical family but had not attended college herself, instead finding a way into science through being a part-time secretarial assistant to Francis Galton.[23] Some of these women wrote or co-wrote papers published in the *Proceedings* and *Philosophical Transactions of the Royal Society of London*, as outlined in the following chapter; some of them were also members of the London Mathematical Society.

The London Mathematical Society

In his 1890 presidential address to the British Association for the Advancement of Science, J.W.L. Glaisher stressed the vital role that the London Mathematical Society (LMS) played in keeping the flame of mathematical enquiry alive.[24] The Society was then just twenty-five years old, but already it was acknowledged as a leading mathematical organisation, not just in Britain, but within Europe too. For any individual seeking to stay up-to-date with research mathematics, membership of this elite learned body was extremely useful. How receptive was this mathematical community to female involvement? The LMS had been established in 1865 by Augustus de Morgan (Ada Lovelace's private tutor) who was concerned to establish a convivial society to which mathematics could be presented and discussed. Although gender was not specified as a factor in membership, in its early years the Society met in the rooms of the Chemical Society (an organisation that was strongly resistant to female participation) at Burlington House, which was also the home of the Royal

Society. It would have been difficult at this time for women to access this 'male' space and, by extension, the LMS itself.

In the early 1870s the Society moved to its own premises at Albermarle Street. The aims and practices of the LMS were restated in 1889 and its membership criteria spelt out: the Society was constituted for the promotion and extension of mathematical knowledge, consisted of ordinary and honorary members, and required each potential member to be proposed and recommended by at least three ordinary members prior to election by a majority. By this time, the LMS had four female fellows: Charlotte Angas Scott had joined in 1881, as had American Christine Ladd-Franklin who had connections to the Society through J.J. Sylvester of Johns Hopkins, a past president of the LMS. Sophie Bryant became a member a year later in 1882. At that time Bryant was teaching mathematics at the North London Collegiate School while researching her doctorate which was awarded by the University of London in 1884. She is generally held to have been the first woman to receive this honour in England, gaining a DSc in Psychology, Logic and Ethics. Theodora Margaret Meyer joined the LMS in 1888, in the same year that she was appointed lecturer at Girton College. Meyer served as director of mathematical studies at Girton 1903–1918, after which she taught at University College London. During World War One she devoted herself to war work, carrying out calculations connected with the construction of aircraft.[25]

In order to illustrate the unique nature of the membership of the LMS as compared to other learned scientific societies, it is useful to analyse the records for one particular year. In November 1903, as well as thirteen honorary members, the Society laid claim to 361 ordinary members of which thirteen (nearly 5%) were women. Approaching 13% of the membership were Fellows of the Royal Society, forty-six members in total. Although the largest occupation group was university teachers, comprising 116 or nearly one third of the membership (including three women) by no means were all LMS fellows mathematicians in any 'professional' sense (this was not surprising as Cambridge, for example, turned out far more wranglers than could possibly hope to achieve a job in academia). Forty school teachers comprised the next largest group, followed by ministers of the Church (nineteen) and barristers/JPs (thirteen). Membership also included civil engineers, a naval instructor, an MP and a mathematical editor and reviewer for *The Times* (Constance Isabella Marks). Forty-eight (just over 13%) listed no occupation at all, among them the largest group of women (seven) including Grace. Even the category of university teachers was not dedicated solely to mathematics but included teachers in related disciplines such as astronomy and physics; a large proportion of overseas fellows were also represented in this group.

Given the occupational (and geographic) fragmentation of the fellowship, it is clear that the LMS did not function as a pressure group for any particular professional interests. At a time when other learned societies were refining their membership criteria in order to impose qualification bars, raise professional status, encourage patronage, and gain power and influence in the modern, industrial world, the LMS retained many of the attributes of a gentler age. One reason for this relative stability was that ways of doing mathematics had not changed significantly over the years: it remained a discipline that was attached to the university, it had not seen a rush from the personal study to new spaces of work such as the laboratory, neither did it require new tools or modern instrumentation. Moreover, no qualifications were stipulated as a necessary requirement for an LMS member and admission still depended on the recommendation and vote of existing members. Although the majority of members would have sat the Cambridge tripos or have achieved a mathematical degree elsewhere this was not an indispensable requirement. For example, Mrs. Alicia Boole Stott had enjoyed no formal mathematical education at all yet was still elected to the Society.[26]

The liberality of the LMS's election criteria can be contrasted to the Royal Society which had made several reforms to its fellowship selection procedures during the mid-nineteenth century designed specifically to exclude 'amateurs'. By end-of-century, the Royal Society had consolidated its position as a consultative authority on science issues and its fellows were frequent recipients of government patronage. Many fellows of the Royal Society were strongly resistant to the election of female members, not least as the presence of women among their ranks was perceived as a possible threat to prestige and professional status. Similar fears were also central to the hostility to women which manifested itself within the Chemical Society, a body which had been formed by chemists in 1877 expressly in the interests of their profession. The Royal Astronomical Society was similarly jealous of its professional status and influence and many of its fellows felt that this could be questioned if women were admitted. When Elizabeth Iris Pogson Kent of Madras University was nominated as an ordinary fellow in 1886, and the Society's legal adviser found no bar to her election, the governing council sought another opinion as a matter of urgency. The second lawyer concluded that those who had written the Royal Astronomical Society charter had not intended to allow women and, as a result, Kent's nomination was withdrawn.[27] At a time when the amateur-professional distinction was seen as increasingly important to a learned association's status, and when the Royal Society, Royal Astronomical Society and others (see Chapter 7) were taking steps to formalise women's ineligibility, the acceptance of women into the LMS can be seen as significant.

Despite the visibility of female members, the actual number of women in the LMS was small in real terms (as Table 6.1 shows) and this, in part, reflected the Cambridge dominance of the Society. In 1903 at least nine of the thirteen female fellows had 'graduated' from the Cambridge mathematics tripos. The other women were Alicia Boole Stott, an independent scholar, Constance Isabella Marks, a mathematical editor with a degree from University College London, and Lilian Janie Whitley, a lecturer in mathematics at Westfield College. But it was not just among women that Cambridge had a strong showing. Cambridge was at the heart of British mathematics at this time and, reflecting this, the presidency and council of the LMS was heavily biased to the University in the two decades surrounding 1900 with at least eight of the ten presidents (each holding office for two years) being products of the Cambridge mathematics tripos. The character of the LMS at the end of the nineteenth century was conveyed in the opening remarks to the Society of one president who stressed that he had 'the honour to address gentlemen who, for the most part, have benefited from the congenial associations and scientific air of the ancient universities ...'.[28] Women were accepted with courtesy by the gentleman members of the LMS 'club', but their participation was not of an equal nature. Women did not gain a presence on the council until the appointment of Hilda Hudson as an ordinary member in 1917; the first woman president was Mary Cartwright who served between 1961 and 1963.[29] Women did not win Society medals, the first female recipient of the LMS De Morgan medal was (again) Mary Cartwright in 1968. William Young won this prestigious medal in 1917, awarded in his name alone despite his wife's key contribution to his papers.

Women were, however, active to a certain extent in presenting papers before the Society and publishing in the *Proceedings*. Sophie Bryant achieved a first with a paper on the geometrical form of cell structures which was read to the Society in 1885 and subsequently published in full. Two years later she read another paper on a different subject, but this time did not publish. Bryant was followed by Frances Hardcastle and Mildred Emily Barwell, both Girton 'graduates' who had passed the specialist part II of the Cambridge mathematics tripos in 1892 and 1896 respectively. Bryant and Barwell (at least) had read and discussed their papers before the Society in person as these discussions are recorded in

Table 6.1 Women members of the London Mathematical Society (LMS) to 1914

NO	NAME & LMS ELECTION DATE	UNIVERSITY	CAREER	PAPERS IN *LMS. PROC.* TO 1914	OTHER MATHEMATICAL PAPERS IN ENGLISH LANG. JOURNALS TO 1914
1	Barwell, Mildred Emily 1898	Cambridge MT (Girton) MA (Dublin)	Maths lecturer (Girton); Maths mistress (Dulwich College; Alexandra College, Dublin)	1897	
2	Bennett, Agnes Elizabeth 1903	Cambridge MT (Girton)	Maths mistress and administrator, North London Collegiate School		
3	Bryant, Sophie 1882	BSc, DSc (London)	Headmistress, North London Collegiate School	1884/5	
4	Cave-Browne-Cave, Beatrice Mabel 1900	Cambridge MT (Girton)	Research stud., Girton; High school teacher; statistician at Biometrics Laboratory (UCL)		*Biometrika** (1914)
5	Cave-Brown-Cave, Frances Evelyn 1900	Cambridge MT (Girton); MA (Dublin)	Maths tutor and Dir. of Maths studies, Girton; statistician at Biometrics Laboratory (UCL)		*RS Proc.* (1904*, 1904/5)
6	Fawcett, Philippa Garrett 1891	Cambridge MT (Newnham)	Fellow UCL; Fellow & tutor, Newnham; Educational Administrator		*QJPAM* (1893); *RS Proc.* (1894)

Table 6.1 Women members of the London Mathematical Society (LMS) to 1914 – *continued*

NO	NAME & LMS ELECTION DATE	UNIVERSITY	CAREER	PAPERS IN *LMS. PROC.* TO 1914	OTHER MATHEMATICAL PAPERS IN ENGLISH LANG. JOURNALS TO 1914
7	Franklin, Christine Ladd 1881	DSc (Johns Hopkins)	Research stud. Johns Hopkins; occasional Maths tutor Johns Hopkins, Columbia and Chicago Universities		*AJM* (1880, 1880)
8	Hardcastle, Frances 1896	Cambridge MT (Girton)	Research stud. Bryn Mawr, Chicago and Girton	1897	*Bull. AMS* (1898); *BA Reps.* (1900, 1902/4)
9	Hudson, Hilda Phoebe 1906	Cambridge MT (Newnham) ScD (Dublin)	Maths tutor and research fellow, Newnham; research student Berlin and Bryn Mawr; Maths tutor West Ham Tech. Institute; part-time assistant, Air Ministry; technical assist. Parnell Aeronautics	1911, 1912, 1913	*AJM* (1912, 1913); *Int. Congress. Math. Rep.* (1913); *QJPAM* (1913)
10	Lees, Edith 1894		Based in London		
11	Long, Marjorie 1910	Cambridge MT (Girton)	Maths tutor, Bedford College	1911	
12	Maddison, Ada I. 1897	Cambridge MT (Girton); BSc (London); PhD (Bryn Mawr)	Research stud., Göttingen; Reader in Maths and administrator, Bryn Mawr		*QJPAM* (1893; 1896); *Bull. AMS* (1896; 1897)

Table 6.1   Women members of the London Mathematical Society (LMS) to
1914 – *continued*

NO	NAME & LMS ELECTION DATE	UNIVERSITY	CAREER	PAPERS IN *LMS*. PROC. TO 1914	OTHER MATHEMATICAL PAPERS IN ENGLISH LANG. JOURNALS TO 1914
13	Marks, Constance Isabella 1903	BA (UCL)	Maths Editor and Reviewer, *Educational Times*		
14	Meyer, Margaret Theodora 1888	Cambridge MT (Girton)	Maths tutor and Dir. of Maths studies, Girton; part-time tech. assist., Air Ministry		
15	Scott, Charlotte Angas 1881	Cambridge MT (Girton); BSc, DSc (London)	Maths tutor, Girton; Prof. Maths, Bryn Mawr		*AJM* (1886, 1899); *AMS Bull.* (1898, 1900, 1905); *AM* (1900, 1906); *Math. Gaz.* (1900); *MM* (1896); *RS Phil. Trans.* (1894); *RS Proc.* (1893); *QJM* (1896, 1899, 1901); *QJPAM* (1896; 1901); *AMS Trans.* (1902)
16	Stott, Alicia 1902	Hon. Degree, (Groningen)	Amateur builder of Maths models		
17	Whitley, Lilian Janie 1902	BA (London)	Maths tutor, Westfield College		
18	Young, Grace Chisholm 1896	Cambridge MT (Girton); DSc (Göttingen)		1910,* 1910*	*QJPAM* (1906; 1909; *1909; *1910; *1913*)

Key: * Co-authored paper; AJM *(American Journal of Mathematics);* AM *(Annals of Mathematics);* AMS Trans. *(Transactions of the American Mathematical Society);* BA Reps. *(Reports of the British Association);* Bull. AMS *(Bulletin of the American Mathematical Society);* LMS. Proc. *(Proceedings of the London Mathematical Society);* Math. Gaz. *(Mathematical Gazette);* MM *(Messenger of Mathematics);* QJM *(Quarterly Journal of Mathematics);* QJPAM *(Quarterly Journal of Pure and Applied Mathematics);* RS Phil. Trans. *(Philosophical Transactions of the Royal Society of London);* RS Proc. *(Proceedings of the Royal Society of London)* the minutes. Barwell read a paper before the Society again two years later while Ernest W. Hobson, who had coached 1890's top wrangler Philippa Fawcett, was in the chair. By turn-of-century then, just three papers authored by women had appeared in the *Proceedings of the LMS*. The following decade of the 1900s is initially puzzling as, despite the number of women fellows, the *Proceedings* published no papers by women during this time. Women were participating from the margins however; in 1907 Alicia Stott attended a meeting of the Society in person and exhibited a collection of mathematical models and there were also three communications received from women during this time. In 1910 Grace and her husband published two joint papers and these were followed in 1911 by a first paper by Hilda Hudson, research fellow at Newnham, and another by Marjorie Long, Assistant in Mathematics at Bedford College, University of London. Hudson contributed two papers again in 1912; in 1913 Miss R.E. Colomb published a joint paper with Harold Hilton, Professor of Mathematics at Bedford College. Add to this another joint paper by Grace and William Young and this makes a total of eight papers by women published in the *Proceedings* up to 1914.

Given the female membership of the LMS at this time, this low level of authorship requires explanation. It is difficult to explain the paucity of female authors by any gendered participation pattern as the actual number of men publishing papers was, in relation to the number of male fellows, similarly low. Because of its prestige and international reputation, the *Proceedings of the LMS* acted as an outlet for a small elite of (mostly) Cambridge-connected mathematicians who repeatedly published in the journal, with up to four or five papers authored by the same person in any one volume. For example, William Henry Young had six substantial papers published in 1910 alone, and he was joined in the same volume by fellow Cambridge mathematicians G.H. Hardy and E.W. Hobson who published two papers each. Most fellows of the LMS, male or female, were passive members comprising an audience for the elite mathematicians whose work they read in the *Proceedings* and discussed at meetings. In this light, women's publication record appears less disappointing, and it is important to remember that Grace's contribution was invisible in the vast number of papers published under her husband's name alone.

Once women had moved on from Cambridge they were less likely to publish in the journal, even if they remained active in research. For example, Charlotte Angas Scott, although a life member of the LMS and a productive paper writer, did not contribute to the Society's *Proceedings*. Many of her papers were published in the *American Journal of Mathematics* of which she became co-editor in 1899. Scott also published in the *Transactions of the Royal Society of London* in 1893 and 1894 and in other British and foreign journals. Hilda Hudson, on the other hand, remained at Cambridge and was a presence in the *Proceedings* with three papers between 1910 and 1912 and several more after 1914. Hudson was one of a Cambridge mathematical family, her father had taught at the University and was later professor of mathematics at the University of London. Hudson had attended Newnham College where she was placed equal to the seventh wrangler in the 1903 tripos (her sister Winifred been placed equal eighth in 1900) and in 1904 had achieved a first-class pass in part II. Hudson followed these Cambridge successes with a year at the University of Berlin before returning to Newnham College to take up a post as residential lecturer, later research fellow.[30] It seems that membership of the LMS was desirable for any female (or male) research mathematician, but publication could be elsewhere. Other significant journals for the publication of mathematics at this time were the *Quarterly Journal of Pure and Applied Mathematics*, the *Messenger of Mathematics* and, to a lesser extent, the *Proceedings of the Cambridge Philosophical Society*. The *Proceedings of the Royal Society* were also a possible outlet and this will be examined in the succeeding chapter.

The Edinburgh and American Mathematical Societies

A brief consideration of the Edinburgh Mathematical Society (EMS) and the American Mathematical Society (AMS) is useful for placing the preceding discussion into wider context and assessing the comparative significance of women's participation in the LMS.

The EMS was founded in 1883 and quickly became an important association for, predominantly, the mathematical community in Scotland. From its inception in 1883, membership grew from fifty-one to 236 in 1913. In contrast to the LMS, the impetus behind the establishment and organisation of the EMS in the nineteenth and early-twentieth centuries came from school teachers rather than university lecturers. At the end of the EMS's first year around forty of its fifty-eight members were school teachers and another fifteen had university connections. This bias persisted and of the first ten presidents seven were school teachers while three were university staff.[31] Despite this link to school teaching, and the fact that no restriction was placed on ordinary membership apart from being proposed and seconded, women did not play a prominent role in the EMS to 1914 nor, indeed, were they represented strongly within the membership. The first woman member was school mistress Flora Philip who, when elected in 1886, was the only woman alongside 132 men.[32] Philip remained a member of the EMS for seven years. There were, altogether, sixteen women variously listed as members of the EMS between 1886 and 1914, nine of whom can be identified as school mistresses; at least six women attended the Society's first mathematical colloquium in Edinburgh in 1913. One notable member during these years was Charlotte Angas Scott who joined the Society in 1897; another was Jessie Chrystal Macmillan, suffrage campaigner and one of the first woman barristers.[33] No woman is listed as a contributor to the Proceedings prior to 1914 however, and it was 1965 before the first woman assumed the office of President of the EMS.[34]

The AMS was established originally in 1888 as the New York Mathematical Society and changed its name in 1894 to reflect its wider ambitions. Its founder, Thomas S. Fiske, sought to emulate the 'comradeship' amongst mathematicians that he had experienced as a guest at LMS meetings on a trip to London in the Spring of 1887. In contrast to the lack of formal qualifications required of LMS and EMS members, AMS membership rules excluded mathematics teachers of less than 'collegiate' rank. This would have excluded many women mathematicians teaching in high schools, one of the few career routes open to mathematical women. Despite this, by the mid-1900s women mathematicians have been identified as a 'visible' group in an increasingly important, research-focused, AMS.[35] The Society grew steadily during the first decade of the twentieth century until, in 1912, women comprised 50 out of a membership of 668.[36] Like the LMS, a majority of these women were not productive in the sense of contributing papers but were, nonetheless, part of a learned membership open only to those with research or advanced teaching credentials. These 'mathematically interested' women were complemented by around a dozen more active female members who attended meetings and, occasionally, contributed papers. Among this group Charlotte Angas Scott was the most prominent, having twelve papers read before the Society and nineteen articles published in the Proceedings. Ida May Schottenfels was the next most active female contributor with seventeen papers read and three published.[37]

The pattern of women's participation in the AMS is broadly similar to the LMS, although the numbers of women are larger due to the greater opportunities available in America for women to gain an advanced mathematical education. Indeed, the AMS noted in its register for 1912 that it enjoyed a larger female membership than comparable European societies, remarking that the German mathematical society had only five female members out of a total of 759 'and only one of these is a German woman, while three of them are Americans and the remaining one is a Russian'.[38] (This was largely due to Germany's reticence to admit its own women to university as discussed in Chapter 2.) However, unlike the LMS, the AMS did appoint a woman to it council before 1900. Charlotte Angas Scott, who had been amongst the first six women to join the AMS in 1891, served on council from 1894, becoming vice president in 1905.

The Cambridge Philosophical Society

At the inaugural meeting of the LMS in 1865, De Morgan spelt out who its main rivals would be and why the formation of this new society was necessary. The Royal Society, he argued, was 'too wide' and the Cambridge Philosophical Society 'too local'; this, he implied, left a gap for a mathematical society to cater for the specific needs of mathematicians in Cambridge, the universities and beyond.[39] By the time De Morgan was speaking, the Cambridge Philosophical Society already enjoyed a long history at the University having been founded in 1819. Its *Proceedings*, which embraced the mathematical and physical sciences, had been published since 1843. Membership of the Society was restricted to Cambridge fellows therefore women were ineligible, but in 1876 new articles of membership had been formulated to allow non-Cambridge graduates to become associates and attend meetings. A history of the Cambridge Philosophical Society, published in 1969, is somewhat defensive regarding the position of women within these new rules and claims that women could be admitted as associates and that, by 1914, they had been 'for some years as a matter of routine'.[40] However, from an analysis of the Society's *Proceedings* in which membership lists were regularly published, there seems evidence to question this claim. For example, in 1907 the Society had 334 fellows, nine associates, twenty council members and thirty-seven honorary members – all male. No women are listed as being admitted until 1914 when Marie Curie was granted an honorary membership. Prior to awarding her this honour, the Society had taken advice as to its legality and were reassured that since honorary fellows had no rights, the Society's bye-laws would not be infringed.[41]

No mathematical papers by women appear in the *Proceedings of the Cambridge Philosophical Society* up to 1914, around which time the journal slowly came to be dominated by the biological and physical sciences. Papers authored or co-authored by women in these disciplines, mostly from William Bateson's research group at Newnham or the few female researchers at the Cavendish Laboratory, begin appearing from 1890 onwards. By 1914 twenty-five papers authored or co-authored by women had been published, prior to this a smaller number of communications by women had been noted. All of these papers had been communicated by a male fellow so it is difficult to ascertain if women were ever actually present in person. However, the Society held its meetings at various locations throughout the University and some of these venues were more welcoming to women than others. Whereas women could be a regular presence at the Cavendish Laboratory or botany school, as a general rule they were not admitted into the University chemistry laboratory (until 1909), the medical schools or the comparative anatomy lecture room – all venues used by the Cambridge Philosophical Society to hold their meetings.

Publishing opportunities

A more viable alternative for mathematical women seeking publication was the *Quarterly Journal of Pure and Applied Mathematics* (*QJPAM*) which was not connected to any learned society. Although founded in 1857, from the early 1870s it had been published by Longmans publishers. The first editorial team included Cambridge's Arthur Cayley and this strong Cambridge connection continued with, for example, J.W.L. Glaisher and A.R. Forsyth on the editorial team from 1895. There was also a considerable overlap of personnel with the Cambridge-dominated editorial board of the *Proceedings of the LMS*. An examination of papers in the *QJPAM* reveals twelve papers by women between 1893 and 1914, including four papers by Grace published jointly with her husband and one published under her name alone. Other female contributors were Charlotte Angas Scott, Isabel Maddison, Hilda Hudson and American Florence E. Allen. If the *QJPAM* contained a Cambridge bias in its editorial team, so did the *Messenger of Mathematics*, despite its founding as a journal to encourage scholarship from students new to mathematical research at the universities of Oxford, Cambridge and Dublin. By the beginning of the second series in 1872, the *Messenger* was inviting scholarship from the 'foremost mathematicians of the age' as well as seeking to be a 'stimulus to research in junior students', emphasising that contributions were not restricted to the universities but could come from any source. Despite this, Cambridge scholars dominated this first collected volume of the monthly journal with twelve papers by Cayley, Glaisher and Routh alone. This Cambridge dominance continued well into the next century, with J.W.L. Glaisher serving as editor from 1887 to the outbreak of World War One. Glaisher also edited *QJPAM* and served as president of the LMS 1884–1886. William Henry Young was a regular contributor here, but women were not well represented with just two small contributions up to 1914.

Women did make a significant contribution to the *Educational Times* however, a publication that differed from others under consideration by being primarily dedicated to problem-solving. The *Educational Times* was edited from 1902–1918 by LMS member Constance Isabella Marks. This monthly journal, established in 1847 and published from 1861–1947 as *The Educational Times and Journal of the College of Preceptors*, carried a popular section devoted to mathematical problems and their solutions as well as notices of scholarships, teaching vacancies and other educational subject matter. Tattersall and McMurran have found that one third of all articles submitted to the *Educational Times* mathematical department under Marks' editorship were by women. The most active period for female contributors was the last decade of the nineteenth century when the majority of female contributors were 'graduates' of Girton and Newnham Colleges. Hertha Ayrton made the largest contribution, forwarding 117 submissions between 1881 and 1899; other women submitting material included Philippa Fawcett, Isabel Maddison, Frances Cave-Brown-Cave, Margaret Meyer and Hilda Hudson – all (except Hertha) members of the LMS. This journal was not a peripheral or specifically amateur outlet; other mathematicians to appear on its pages included Felix Klein, Augustus de Morgan, J.J. Sylvester, Arthur Cayley, James Clerk Maxwell and others. The *Educational Times* was also the outlet for the first mathematical publications by G.H. Hardy and Bertrand Russell.[42]

The above discussion suggests that mathematics in England in the decades surrounding 1900 comprised a small community of amateur and university mathematicians. This community was dominated by a narrow elite of mainly Cambridge scholars who were active in publishing papers, serving as editors on various journals, and sitting on the council of the LMS. Mathematics continued to be attached to the university and did not professionalise or modernise its working practices as did other scientific disciplines during this time. As a result, women were able to move from university mathematics to publishing research papers and becoming members of the LMS without facing the same level of hostility than was the case in the physical sciences. This was in part because of their prior university connections (the majority of women active in mathematics were 'graduates' of Girton and Newnham) and partly because mathematics retained a strong 'amateur' flavour. Because of this, women mathematicians were not perceived as a threat to mathematics' professional status, nor were they in competition with men for jobs as they were limited to teaching at the new colleges for women in the UK and in the States. The numbers of women becoming members of the LMS and publishing research papers is small in real terms but can be seen as historically significant in percentage terms: the relatively high proportion of female members of the LMS in the early 1900s was matched again only in the 1970s. A similar pattern of participation of women in mathematical research has been identifed in America by Green and LaDuke and in Germany by Renate Tobies.[43]

Mathematical congresses

Another place for advanced mathematical women to interact with the wider mathematical community was at the regular International Congress of Mathematics. Following the initial meeting in 1893, these events became an important venue for networking and were a showcase for some of the most respected mathematicians of the day. Tracing women's involvement is difficult as most of the women who attended congresses at this time were there as the wives or daughters of male mathematicians (although of course some may have been there out of an interest in mathematics too). It seems certain however that women's direct involvement in these conferences was limited; a few female mathematicians can be identified as delegates, although only two papers written by women were read, one in 1908 and one in 1912.

The first Mathematical Congress was held in Chicago in 1893 to coincide with the establishment of the new University there. The event was opened by Felix Klein who stayed on after its close to give a series of influential lectures at Northwestern University, known as the 'Evanston Colloquium'. Among the forty-five names on the official register are four American women: Professor Achsah M. Ely of Vassar College and graduate students Charlotte C. Barnum, Ida Schottenfels and Mary (May) Winston.[44] At this time Winston had already applied to study at Göttingen and had travelled to Evanston on the invitation of Felix Klein who interviewed her while she was there. Winston's papers reveal that Christine Ladd-Franklin also attended Klein's lectures, plus there is a suggestion that Ruth Gentry may have done so too, although it seems the latter two were not registered for the Congress.[45] At the following Congress, Zurich in 1897, four women can be identified as independent participants, including the first woman to graduate in mathematics in Italy, Iginia Massarini.[46] Among the some 229 delegates at the next event, Paris in 1900, were Charlotte Angas Scott, May Winston and around six other women who may have been there for the mathematics.[47] Although neither Scott or Winston gave a paper, both made a contribution: Angas

Scott reported on the Congress for the AMS, and Winston translated Hilbert's list of 123 mathematical problems, which he had presented before the Congress, for the *Bulletin of the AMS* in 1902. (Hilbert's famous 'problems' were highly influential and continue to set an agenda for mathematical research today.)

The next conference took place in Heidelberg in 1904; here it is difficult to identify firmly any women mathematicians taking part in the proceedings, although approaching sixty women's names are listed as 'accompanying persons'. This distinction was not always a real one however. At the following event (Rome 1908) Max Noëther brought his daughter Emmy as a guest. Noëther had gained her doctorate in mathematics from Erlangen the previous year and was currently assisting her father, a professor of mathematics there, in an unofficial capacity. Then twenty-six years old, she was just beginning to embark on the research, including contributions to the mathematical foundations of relativity theory, which was to be so influential and win her the respect of her obituarist Albert Einstein.[48] This Congress is also notable for the first paper by a woman being read, although not by herself alas. A male delegate presented the young Italian researcher Laura Pisati's work to the Congress as she had died earlier that year.[49]

The 1912 Cambridge International Congress of Mathematics was notable for the large number of delegates (around 700) and for a much greater feminine presence than usual. Mostly wives and daughters of male mathematicians, all 152 of these women were housed in the Cambridge women's colleges and were looked after by a 'ladies committee' which ran excursions, tea parties and other activities. However, analysis of the delegate list reveals that up to forty-five were there independently and it is likely that these women were mathematicians or mathematically-interested. Many of these were graduates of the women's colleges and/or mathematics mistresses from girls' high schools such as Cheltenham Ladies College. Those present included botanist Agnes Arber, Frances Cave-Browne-Cave, Elizabeth Cowley from Vassar College, Margaret Meyer and Hilda Hudson. Hudson was distinguished from her mathematical sisters by giving a paper before this illustrious mathematical audience, presenting in 'Section II: Geometry'.[50] What made mathematics an attractive option for these women in the decades around 1900, and were there other factors influencing women's participation in this community?

A female affinity with mathematics?

In 1898 Professor E.B. Elliott, President of the LMS, felt it necessary to make a plea in his outgoing address for the importance of mathematics and mathematical training. Elliott was dismayed at the growing loss of students to the natural sciences tripos and the consequent 'numerical weakening' of mathematical candidates which, he argued, 'should be striven against with all our energies'. His defence of his discipline, which was made just two years before the start of the twentieth century, centred not on the utility of mathematics or any service that it could render to society, instead it referred back to the older tradition of mathematics as a crucial part of a 'liberal education' designed to encourage moral and intellectual capabilities:

> We believe a sound mathematical training to be of such value in mental development … (that) … life-long application to mathematical investigation, on the part of those who prove to be qualified for it, is a noble use of the human mind, a devotion to the pursuit of truth in a field where the human intellect is on sure ground.[51]

As discussed with reference to women's mathematical training at Girton in Chapter 1, the purpose of a liberal education was not to prepare the student for any specific role in the world, or to provide any kind of professional training, instead the individual and his or her personal development (in terms of self-discipline and character formation) was the priority. In the closing decades of the nineteenth century, the growth of physical mathematics and technology, with their emphasis on learning about the world and creating new methods of controlling it, had eroded the argument that mathematics, as part of a liberal education, was best pursued for its own sake and to aid the mental cultivation of the individual. Cambridge men were now deserting the mathematics tripos for the natural sciences with, for Elliott and his colleagues, alarming speed. By end-of-century the notion of a liberal education was retained by pure mathematicians, in the absence of any justification of their subject on the grounds of utility, as one way to legitimise their discipline.

The ideal of a liberal education was also used by advocates of higher education for women as a powerful argument to promote their cause. As James Orton wrote in *The Liberal Education of Women*, mathematics (and the classics) 'tend to enlarge and strengthen the moral and intellectual powers, and this is necessary *sub modo* for women as well as for men'.[52] His point was that higher education would help women fulfil their potential as wives and mothers, rather than seduce women away from these roles as feared by opponents of the new experiment in female education. In short, a liberal education was best suited for women and mathematics, with its long association with these ideals, was an appropriate and comfortable choice. Moreover, mathematics was less likely than laboratory work to compromise a woman's femininity or to threaten feminine propriety with unwholesome knowledge of the world. On the contrary, with its traditional connection to the development of moral character and good sense, mathematics could be the ideal tool to prepare women for their role as society's – and men's – superior conscience.

The changes in the conception and configuration of mathematics that gave the discipline an affinity with women did not rest solely on its advantages in imparting a liberal education; as discussed above women's choices were also limited by their schooling which often neglected applied mathematics and science. As has already been outlined in Chapter 4, pure mathematics became more 'feminised' in its use of terminology at turn-of-century, especially in relation to Göttingen-style analysis. The language of 'elegant' theorems or 'beautiful' proofs contrasted sharply with that often applied to experimental science which could call for 'courage', 'perseverance' or 'iron nerve' (as discussed in Chapter 5). In this way, the differing language and metaphors of feminised pure mathematics and robust, masculinised science, worked to naturalise each discipline's separate practices of inclusion or exclusion. Arguably these factors combined to make women feel more 'at home' with pure than applied mathematics, and data on early women PhDs tends to show a preference for topics in pure mathematics, especially geometry. This holds true for Germany[53] and America.[54] Of the early British women mathematicians, Charlotte Angas Scott, Grace Chisholm Young, Isabel Maddison and Hilda Hudson all wrote their doctoral theses on algebraic or geometric topics.

At the same time that women were gaining a foothold on advanced mathematics in England, so there developed a tendency for intellectual work in general to be viewed as feminine. This contrast was made more acute by a new trend for rugged, athletic manliness, with anti-intellectual associations, which began to gain ground.[55] This was reflected at Cambridge by a dilution of the connection between mathematics and rigorous bodily training, a flow of men from the mathematics to the natural sciences tripos, and the introduction of a new end-of-century stereotype: the anti-intellectual, sports-motivated 'university hearty'.[56] That 'brain work' could be viewed as particularly feminine was implied by James Orton in a passage quoted in Chapter 3. Orton argued that, in many respects, the sedentary student life was more natural to women than to men; men were now increasingly at the vanguard of physical science and engaged in leading the way in the growth of new technology.[57] In this way, Orton and others upended older ideas about women's incapacity for rational thought and argued that mathematics, art, classics and similar scholarly pursuits could be the natural preserve of women.

The pure and the practical – significations beyond mathematics

Central to views about the appropriateness of various disciplines to either sex was the idea that there is a fundamental difference between pure mathematics and practical mathematics and science. It has been a scholarly tradition to classify women mathematicians with women scientists but, as this study shows, there was a marked contrast in the material practices of science and mathematics and, at varying historical periods, key differences in what each discipline signified. This was especially so at the end-of-century when the breakdown of the earlier tradition of 'mixed mathematics' facilitated the rise of the new 'pure mathematics' with its contrasting aesthetics and sensibilities. As a result, subsuming women mathematicians with women scientists runs the risk of losing sight of these differences – differences that can be key to an understanding of women's participation. There is no doubt that a clear separation was made at this time between pure and more practically-oriented mathematics; indeed the difference between the two was seen as important enough to provide the theme for J.W.L. Glaisher's 1890 Presidential Address to the British Association for the Advancement of Science. Glaisher's speech was a rallying cry for pure mathematics which he saw as under threat by the accelerating growth of 'physical mathematics'. Although Glaisher did not seek to undervalue 'those branches of mathematics which we owe to the mathematical necessities of physical inquiry' he followed this by creating a hierarchy between the two which clearly privileged the pure:

> But it always appears to me that there is a certain perfection, and also a certain luxuriance and exuberance, in the pure sciences which have resulted from the unaided, and I might also say inspired, genius of the greatest mathematicians which is conspicuously absent from most of the investigations which have had their origin in the attempt to forge the weapons required for research in the less abstract sciences.[58]

Glaisher's arguments parallel the debates going on in Germany at this time and presage the remarks of English mathematician G.H. Hardy who contrasted 'uncontaminated' pure mathematics with the ugly utility of the applied. In England as in Germany, pure mathematics was fighting for its place in the face of mathematical physics and the practical sciences which were seen as bringing the important benefits of utility to the *real* world. At Cambridge this tradition of 'mixed' mathematics had produced great figures of mathematical physics such as James Clerk Maxwell and Lord Rayleigh and, in the face of this pre-eminence, pure mathematicians such as Glaisher and Arthur Cayley were moved to justify their discipline.[59] Glaisher used familiar arguments to legitimise and add moral value to his discipline: that pure mathematics was the only way to reveal certain truth and that pure mathematics, dependent as it was on the creativity and power of individual minds, was the natural home of genius. He finished by broadening his remarks to include society in general, implying a lament for the old established order and warning that the 'search after abstract truth for its own sake, without the smallest thought of practical application or return in any form ... are signs of the vitality of a people, which are among the first to disappear when decay begins'.

Glaisher's conception of pure mathematics was unashamedly elitist. Only special individuals were equipped with the necessary mathematical aptitude and even they required years of study to learn the language and acquire the necessary expertise. This combined emphasis on the power of the individual intellect (a mathematician working in inspired isolation with his brain, not his hands) and a language requiring special skill, imposed 'of necessity ... narrow limitations on the members of our audience'. There was also a class dimension here: Glaisher's words rehearsed the long-standing recognition of the 'higher' moral worth of 'brain-work' over 'hand-work', and by claiming pure mathematicians as a 'natural born' elite he effectively discounted the possibility of infiltration via education by any 'lower' intellectual ranks of society. Pure mathematicians were unanimous in their fears that their specialism was losing out to 'physical mathematics' and, as a result, they were more inclined to accept elite women with the necessary expertise in order to bolster their numbers and strength. However this acceptance was tempered by an admiration for, as Glaisher put it, the 'inspired genius of the greatest mathematicians'. This referred back to a romantic ideology of genius with which Victorian thinkers were fascinated. According to this tradition, the genius was born with a natural gift endowed by Nature – a condition which was inextricably linked to the masculine. Any attempt to apply genius to women, given the gendered history and derivation of the concept, inevitably invoked a tension, as discussed in Chapter 2. In

England as in Germany, romantic ideology, with its feminine colouring yet intrinsically masculine core, privileged the male mathematical mind and ensured that gender hierarchies remained in place even as women were welcomed into the mathematical community.[60]

Conclusion

Historians have pointed to the long-standing connection between masculinity and rationality on the one hand, and femininity and the emotions on the other, exploring how these dualities have led to the exclusion of women from mathematics (and the sciences). Mathematics was believed 'unnatural' for women as it required a level of abstract thinking that was beyond feminine capabilities. Without seeking to deny these connections, it is important to recognise the historical nature of mathematics and gender, and how relations between the two, including the intellectual resources drawn upon to propagate them, change and develop over time.

At the end of the nineteenth century the nature and meanings of mathematics and science were being contested just as the movement for suffrage and higher education were questioning what were acceptable roles for women. Physical mathematics and laboratory science became increasingly professionalised and represented themselves as manly pursuits, fully engaged in the world and capable of 'mastering' it. These disciplines were 'manned' by the new scientific professional – professionals who were concerned that any overt hint of femininity could undermine their growing status (as will be demonstrated with a contextualisation of Hertha Ayrton's career in the following chapter). In contrast, pure mathematics fled from the world into abstraction, legitimised itself through romantic notions of genius and the mystery of 'great intellects', and, as Glaisher implied, clung to older ideas of a natural-born elite whose altruistic talents guarded against 'decay'. At this time of fluidity and negotiation, as mathematics became less conflicted with femininity, and pure mathematicians' numbers decreased, women were accepted into the shrinking mathematical community (albeit up to a point) as pure mathematics became configured as a suitable pursuit for the new, elite, educated gentlewoman.

7
Bodies of Controversy: Women and the Royal Society of London

Narratives of Hertha Ayrton have tended to portray her life as a series of 'break-throughs' and 'firsts'. Scholars have recounted how, for example, when John Perry read Hertha's paper on the electric arc before a meeting of the Royal Society in 1901 it marked the first time that a paper by a woman had ever been read. Similarly, Hertha is credited as taking 'a great stride for all women in the sciences' by becoming the first woman to read her own paper before the Royal Society on June 16 1904.[1] These biographic reclamations of Hertha have played a vital role in making her achievements visible in a culture that still tends to write the history of science from a masculine perspective. However a reassessment of her career using varied sources poses questions for any understanding of Hertha as a lone.pioneer.

A first source of information about Hertha Ayrton is a memoir written by her friend, Evelyn Sharp, who notes that the Royal Society would not allow a woman to speak before them so John Perry, FRS, read her paper on her behalf. Sharp goes on to relate that, three years later, Hertha herself read a paper before the Royal Society 'in person, the first woman to do so ... accompanied by experiments with sand and water in glass troughs'.[2] This 'break-through' account is typical of a memoir which offers only a straight-forwardly narrative account of Hertha's life and which is imbued with feminism and heroism. However, although long and comprehensive, Sharp's text is at times at odds with other source material and is an uncritical celebration of her friend's life as 'physicist, suffragette, democrat and humanitarian'.[3]

The idea of a 'memoir' had originated from a request made by Hertha's daughter, Barbara Ayrton Gould,[4] shortly after her mother's death. In accepting the project, there is no doubt that Sharp intended to write for a particular audience or that she had a specific feminist message to advance.[5] Sharp was a journalist, writing for the *Manchester Guardian* and other papers, and a committed suffrage campaigner. She had met Hertha in 1906 when they both became members of Mrs. Pankhurst's new Women's Social and Political Union. Sharp was imprisoned in Holloway on two occasions due to her militant activities; in 1912 she broke away from the Pankhursts to help found the United Suffragists and edit its journal *Votes for Women*. Sharp's strategy in her journalism and elsewhere was to convey suffrage ideas to the wider public[6] and, in Hertha Ayrton, she was presented with an ideal vehicle with which to argue for women's intellectual equality in a 'male' sphere and to justify the suffrage campaign. (Sharp's biography was published in 1926, two years before women were eligible to vote on the same terms as men.)[7] With this agenda a priority, Sharp's work offers anything but a critical analysis and must be used with caution in the absence of further contextual evidence.

This chapter builds on existing accounts of Hertha's life by attempting a more analytic picture of her achievements. Her experiences are contextualised within the science and gender politics of the time, demonstrating that the difficulties facing women in science were different from those experienced by women mathematicians. An analysis of the procedures, politics and motivations of the Royal Society – the context within which many of Hertha's successes were made – reveals a more complex story than just one woman 'breaking through' entrenched institutional misogyny. The discussion will commence with a consideration of women's relationship with the Royal Society, assessed from a number of perspectives including papers published by women and women's involvement in Royal Society conversaziones. Close attention is paid to the Society's award of the Hughes Medal to Hertha in 1906 and to the negotiations surrounding her nomination to a fellowship in 1902. It will be illustrated that Hertha's failure to gain entry to the Royal Society had a direct impact on her credibility and on her work. Yet the Royal Society was not wholly anti-women: assessing and arbitrating on the work of the increasing number of female scientists emerging from the expanding colleges for women was essential to the Society in its aim to remain the 'gatekeeper' and 'judge' of elite science.

The Royal Society and female 'firsts'

Hertha Ayrton was not the first woman to have her work read before the Royal Society. Mathematician Mary Somerville's paper, 'On the Magnetizing Power of the More Refrangible Solar Rays', was presented quietly listens, too (the speaker's enemies "[t]alk with [their] harts" yet only "to [them] selves disclose").

What is also striking is that Protestant iconoclasm is motivated by sibling rivalry. Rather than something inspired by frustration, "making fictions" promote it—as if God had punished the idolatrous Israelites with a golden cow:

> For he him self doth praise
> When he his lust doth ease:
> Extolling rav'nous gaine,
> But doth God's self disdaine.
> Nay so proud is his puffed thought,
> That after God he never sought;
> But rather much he fancies this,
> That name of God a fable is. (Psalm 10 ll. 9–16)

Recalling his criticisms in the *Defence* of love poets who inevitably fail to enchant a lady, Sidney's speaker now chastises poets who convert everything to fiction. The Petrarchan lyric cycle is pointedly rewritten, too, for God flees the lusty poet hunting lies. Another revision of Astrophil's erotic ambition is supplied by Psalm 7:

> Lo, he that first conceav'd a wretched thought,
> And greate with child of mischief travel'd long,
> Now brought a bed, hath brought nought foorth but nought. (ll. 36–39)

Likewise, Psalm 8 echoes Sonnet 31 from *Astrophil and Stella* (where the repeated term "love" only signaled the lover's confinement in a subjective prison), but now the poet's God escapes this imaginative enthrallment:

> When I upon the heav'ns do look,
> Which all from thee their essence took;
> When Moon and Starrs, my thoughts beholdeth,
> Whose life no life but of thee holdeth:
>
> Then thinck I: Ah, what is this man
> Whom that greate God remember can?
> And what the race, of him descended,
> It should be ought of God attended? (ll. 9–16)

This speaker must therefore turn over more serious literary tasks to his superiors, requesting in Psalm 35 that God "Speake thou for me against wrong speaking foes:" (l. 1). The speaker's tongue can easily lead him astray: "And so I nothing said, I muett stood, / I silence kept, even in the good" (Psalm 39 ll. 5–6). Apparently this is a self that works best in secret, unaware and unknown to himself, as we see in Psalm 19:

> Who is the man, that ever can
> His faultes know and acknowledg!
> O Lord, clense me from faultes that be
> Most secret from all knowledg. (ll. 49–52)

* * *

Sidney's fugitive figure unites the legendary psalmist to the fabled Renaissance hero, supplying a myth about poetic power as something at once persuasive and easily ignored, authoritative and undemanding. Mary's psalms do not rescue Philip's speaker from isolation so much as put him on his feet. Maybe this is because Mary recognizes the immanence of sacred and secular worlds, a connection expressed immediately in the dedicatory poem addressed to her brother: "[W]hat is mine / inspired by thee, thy secrett power imprest. / So dar'd my Muse with thine it selfe combine, / as mortall stuffe with that which is divine." This divine access also opens up psychic passages, so that when Mary records the interrelations of passion and pain, in her hands they become contexts, not feelings. We see this in Psalm 58, where sin is rendered as a condition of immaturity or vulnerability, an undeveloped rather than unchosen state. Mary's enemies, she predicts, will melt as the "dishowsed snaile" or "the Embrio, whose vitall band / Breakes er it holdes" (ll. 21–23).[61]

Also striking about the analogy in Psalm 58 is the way genealogy becomes vital and explanatory, even exculpatory, not a hard fact, but a self-renewing premise. This is suggested as well in Mary's outline of the rhythms of subordination and ordination in the very first translation she undertakes, Psalm 44: "Unto thee stand I subjected, / I that did of Jacob spring" (ll. 17–18). Her sacred poise turns poetry into a tool, not a place to collect one's energies but a means to apportion them divinely:

> World-dwellers all give heede to what I saie;
> To all I speake, to rich, poore, high and low;
> Knowledg the subject is my hart conceaves,
> Wisdome the wordes shall from my mouth proceed:
> Which I will measure by melodious eare,
> And ridled speech to tuned harp accord. (Psalm 49 ll. 1–6)

Yet such divine tasks rarely put Mary at the forefront of things: she is a conduit, not a conductor. It is easy to see this difference when we compare her Psalm 63 to Philip's melancholy Psalm 31, where sleeping figures mirror dead ones. In Mary's vision nighttime torpor gives rise to a variety of shapes and possibilities, allowing the poet to freely meditate on the forms of power and powerlessness and muse on the shapes of angels. Yet she quickly passes over these bodily forms because she finds a way to embody heavenly authority, discerning it too must have a clearcut form and a "right right hand":

> ev'n heer I mind thee in my bedd,
> And interrupt my sleepes with nightly thought,
> How thou hast bene the target of my hedd,
> How thy wings shadow hath my safty wrought.
> And, though my body from thy view be brought,
> Yet fixt on thee my loving soule remaines,
> Whose right right hand from falling, me retaines. (Psalm 63 ll. 15–21)

The same movement—from childlike watchfulness to mature, even mathematical reasoning—informs Psalm 139, half epithalamium, half divine-summation:

> Each inmost peece in me is thine:
> While yet I in my mother dwelt,
> All that me cladd
> From thee I hadd. (ll. 43–46)

Often Mary credits her God with the same aesthetic sensibility and delicacy she possesses; He merely is a better artist with similar ambitions. Psalm 139 describes the speaker as grateful for her design: "In brave embrod'ry faire araid" (l. 55). If this artful God actively enriches language and poetry, her poetic promptly traces His handiwork: in Psalm 90 she prays: "Make thou us wise, we wise shall be," "With thy beauty beautify" (ll. 44, 54). Moreover, Mary's psalms make room for a mutuality we have not seen in Philip's work, as if being chosen was like being intimate or being delighted: "How pleasing to my tast!" are God's teachings (Psalm 119 stanza N l. 19): "A plott to please my mind" (Psalm 132 l. 12). Indeed, His "plott" is accessible to sense and sensation equally: "Which though but heard, we know most true to be" (Psalm 78 l. 7). How else would Mary's speaker be able to alliterate sin with knowledge: "For I, alas, acknowledging doe know / My filthie fault, my faultie filthiness" (Psalm 51 ll. 8–9)? How else would Mary's speaker be permitted to take her mother's sins into account and just as fully, knowledgeably absolve her, too:

> My mother, loe! when I began to be,
> Conceaving me, with me did sinne conceave:
> And as with living heate she cherisht me,
> Corruption did like cherishing receave
> But loe, thy love to purest good doth cleave,
> And inward truth which hardlie els discerned,
> My trewand soule in thy hid schoole hath learned.
> (Psalm 51 ll. 15–21)

As Margaret P. Hannay explains, Mary Sidney "characteristically expands images of women's experience whenever they are present or implied in her sources, including the court duties of the aristocratic woman as well as the more universal female experiences of marriage and childbirth."[62] But I think the countess goes beyond this expansion in the *Psalmes*, because female experience becomes the best metaphor for sacred experience and for the creative life. Even when abject, Mary's speaker sees enough to keep things in proportion, realizing: "O God that art more high than I am lowe" (Psalm 56 l. 5). Moreover, Mary's psalms work to find an audience and consolation in a similarly autonomous being, God. In the process, she singlehandedly destroys the social world of the Renaissance lyric her brother Philip reluctantly fashioned. Mary also provides this redefined vocation of poetry with its first success story:

> Thou didst, O Lord, with carefull counting, looke
> On ev'ry jorney I, poore exile, tooke:
> Ev'ry teare from my sad eyes
> Saved in thy bottle lyes,
> These matters are all entred in thy book. (Psalm 56 ll. 21–25)

Again and again we have the powerful suggestion that lyric actually helps the speaker in her autobiographical project, allowing her to be known and be known as a bearer of wisdom. The *Psalmes* allow Mary to rescue poetry from poetics and save her dead brother from a failed earthly life. If he becomes the "pure" "Angell spirit" she describes in her epigraph, she now seems to answer his prayers.

* * *

Mary Herbert appears untouched by the tensions Waller (1974) sees operating in Philip Sidney's poetry, tensions he argues stem from the "complexity in the intellectual history of the Sidney circle" itself, the "tension between Courtier and Christian, between Castiglione and Calvin." But in another context Waller suggests this tension has its advantages:

> The combination of [Petrarchism and Protestantism] produces the peculiar strength of Sidney's poetry, and more generally, those curious Sidneian hybrids, the Protestant Petrarchan love-poetry of the Sidneys, Philip and Robert, Greville, Spenser and the Petrarchan religious lyrics of the Sidneys (this time Philip and Mary) and, eventually, the Sidney's relation, George Herbert.[63]

From the very outset, Petrarch's contribution had been to turn matters of content into principles of form and linguistic tensions into erotic crises. In *Astrophil and Stella*, Philip Sidney challenges Petrarch's transformations or conceits about poetry, sometimes within the space of a single sonnet (the erotic and philosophical reversals in Sonnets 1 and 71 come to mind). This contribution allows Sidney to hold, as one critic notes, "a unique role among his contemporaries both as a model for English poets and as an original voice."[64]

This uniqueness has a price, though, and the alienation can be debilitating: we often see Astrophil unable to position himself firmly, empowered only when others are disabled. In both Astrophil and the erstwhile hero of Sidney's *Defence*, the Italian squire John Pietro Pugliano, Sidney employs the trope of the second player or straw man, the foolish lover or the presumptuous soldier; in the *Psalmes* he describes these figures as jealous idolaters or impious hacks. Sidney adopts this method to beg the self that contrasts with the ruined projects surrounding him. This is a poetry of fame that continually estranges him, and as Greene (1991) suggests, such techniques turn older brothers or poetic antecedents into younger brothers and pale copies. No wonder Philip's model works best when it is suppressed or apprehended from some remove—even its subtleties are too destructive.

For Robert, the poet's isolation is existential, not psychological, brutally physical, not mental. To a degree, of course, Robert has reason to emphasize the physical nature of his isolation, stationed for years at a faraway post his brother had only briefly held. When Robert's long-lost (and nearly forgotten) manuscript was rediscovered it was first ascribed to his brother and then to his uncle (the first earl of Leicester).[65] But Robert's reading of his brother's lonely position prevents him from ever seeing his own isolation as necessary.

Robert's notebook collection includes thirty-five sonnets as well as pastorals, epigrams and elegies, translations of Montemayor's *Diana*, and two Senecan poems (which Philip also translated). Throughout, Robert employs the same conventional themes of courtly love, his collection marked too by the same kind of metrical experimentation found in both siblings' work. But Robert's readers are neither as apt nor as inspired as is Sidney's Stella or Mary's God. Robert's speaker reports in Song 10: "I through a veil saw glory," for "I was part of love's story"; poetry has become another obstacle, and the speaker's expression continually muted or compromised. We are told in Sonnet 30: "I present, absent am; unseen in sight" since "Present not hearkened to, absent forgot" (ll. 4, 14). Even influence gets clotted in Sonnet 34, a reworking of Philip's Sonnet 15, because the milk that sustained Astrophil has now become a deadly poison:

> Whose balms, delaying death, make pains to live,
> That even my wounds my veins with new blood fill
> While beauty's breasts milk to infections give: (Sonnet 34 ll. 10–12)

The solution Robert's poetics repeatedly provide is a speaker who can learn only from his own example:

> Blest be mine eyes, by whom my heart was brought
> To vow to you all my devotions;
> Blest be my heart, by whom mine eyes were taught
> Only to joy in your perfections:
> O only fair, to you I live and move:
> You for yourself, myself for you I love.
> (Sonnet 3 ll. 9–14)

Everything collapses in this picture, even the Petrarchan illusion of two beings attached by voice, eyes, or heart but otherwise entirely removed. Now there is no room to maneuver, no way out. Sonnet 9 echoes this restrictive process with its crippled, crackling iambic: "I yield, I love: to you, than erst, I burn / More hot, more pure; like wood oft warm before, / But to you burnt to dust, can burn no more" (ll. 12–14). This contagion becomes the subject of Sonnet 26:

> Ah dearest limbs, my life's best joy and stay,
> How must I thus let you be cut from me. (ll. 1–2)

Petrarchan language now redounds on the speaker, whose own body— rather than his lady's—comes apart. Certainly one recalls Philip's broken body after the fatal skirmish at Zutphen when the only hope of saving him would have been the amputation of the infected leg. But Robert's poetic distortions are now objects of scrutiny themselves, as if everything the poet muses over instantly gets broken:

> And losing you, myself unuseful see,
> And keeping you, cast life and all away. (ll. 3–4)

> Full of dead gangrenes doth the sick man say
> Whose death of part, health of the rest must be;
> Alas, my love, from no infections free,
> Like law doth give of it or my decay: (ll. 3–8)

Perhaps Philip had buried his work or secreted his voice like a ventriloquist who could make others read and write; but Robert carries out the submersion in a deeper, more deliberate fashion with the logic and the fatality of a deadly disease.

It must be the principles and sentiments of these poems that allow Gary Waller to posit the "unthreatening" figure of Robert (along the same lines, Waller elsewhere presents "the figure of [the] dead . . . [and] increasingly mythologized . . . Sir Philip Sidney").[66] Yet I think Robert actually endangers what he sees, even as he continually aims to fashion a world impervious to his influence or designs. In Pastoral 8 a nymph finds her lovesick shepherd "benighted" and "cold" and tells him: "Thou wer'st as good to talk unto a stone." Still, the shepherd pledges to love the stern nymph, murmuring the hushed threat of a spurned lover or second son:

> The humble shrub whose welfare heaven neglects
> Looks yet to heaven as well as favoured pine.
> Love whom you list: please your own choice, not mine.
> My soul to love and look: naught else affects.

The speaker outlines a new method of artistry and resourcefulness that delegates to others the rewards and fame. In Sonnet 18 he instructs his mistress to "laurel wear in triumph of my bands" (l. 4).

Sonnet 20 makes the same renunciations, although "ere long will a sun rise from the East / In whose clear flames your sparks obscured will be"; "Till then, in sprite those hid beams I adore / And know more stars I see, my night the more" (ll. 11–14). Robert's poetry supplies fictions for the disenfranchised. We see its strained resources in Sonnet 21: "Alas why say you I am rich? when I / Do beg, and begging scant a life sustain:" (ll. 1–2);

"O let me know myself!" (l. 5). And yet a picture of waste is continually revitalized or naturalized as in Sonnet 25:

> You that take pleasure in your cruelty
> And place your health in my infections;
> You that add sorrows to afflictions
> And think your wealth shines in my poverty,

Robert's strict economy of desire is organized much like the one Schwartz describes in her discussion of biblical brothers:

> Since that there is all inequality
> Between my wants and your perfections,
> Between your scorns and my affections,
> Between my bands and your sov'ranity,
>
> O love yourself: be you yourself your care:

He makes himself at once the ghost who abandons earthly things for a place by his brother's side and a Renaissance Cain who abjures the God who "scorns" his "affections."

*   *   *

At work in both Robert's and Mary's poetry is an effort to quell Philip's image's ideological faults as well as an attempt to comprehend their brother's doubts through the unveiling of more stable or coherent worlds. But their work provides additional services, for if influence can simply be a matter of "ghost-writing," it can also involve exorcising a spirit, controlling a phantasm, or simply caring for the dead.[67] Robert places the bulk of literary inheritance, its riches, and its privileges, at his brother's feet, while Mary elevates Philip to a region where he can remain safely silent, finally superannuated. Perhaps her generosity best characterizes the difference between Mary's productive "invisibility," cloistered at Wilton (at least until 1601) and Robert's own exile, banished to his brother's post in the Low Countries for fifteen years.[68] Mary's ability to hide her whereabouts and call attention to her detachment is distinct from the disengagement Robert describes, bereft of an audience and unsure of his voice. Interestingly, some critics claim Robert's language flourishes in his daughter Lady Mary Wroth's sonnets and in her *Urania*, a romance that Josephine Roberts claims mediates "between the ideal realm of retreat and the actual world of court-life."[69] I would argue that both Robert Sidney and Mary Herbert become to the Society by her husband in 1826 and published in the *Philosophical Transactions*.[8] Women's papers were appearing in both the *Philosophical Transactions* and the *Proceedings of the Royal Society* well before Hertha's paper on the electric arc was published in 1901. The earliest paper by a woman has been identified as Ann Whitfield's who wrote on the effects of a thunderstorm in Rickmansworth, Hertfordshire, in 1760.[9] At the end of the eighteenth century astronomer Caroline Herschel published several papers, including one detailing the discovery of three nebulae. This latter paper was highly commended by the Royal Society, which took the rare step of formally announcing Herschel's discoveries by letter to astronomers in Paris and Munich.[10] It would have been difficult for the Society to do this if her discoveries had not been put before the fellows.

Table 7.1 shows that sixty women together published more than 170 papers in the *Proceedings* and the *Philosophical Transactions of the Royal Society of London* between 1880–1914 including fifty papers before 1900. Although numbers are small compared to male authors, contributions by women increased during the final decade of the nineteenth century, not least as students in biological, chemical and physical sciences at the new women's colleges made their presence felt. The new century saw yet another 'rush' by women, including a flurry of papers co-authored by various female researchers and Karl Pearson, Professor of Applied Mathematics and Mechanics at University College London. In the volume of the *Proceedings* in which Hertha Ayrton's abstract on sand ripples was published (vol. 74, 1904–5) there were five papers published by women.[11] Around this time women were increasingly within the pages of other scientific journals too, including the *Proceedings of the London Mathematical Society, Nature, Philosophical Magazine* and the *Proceedings of the Institution of Civil Engineers*.[12] These are modest gains and it is important not to over estimate female participation, however it is clear that Hertha Ayrton was not alone in pushing at the door of the Royal Society. Although for the most part the door held fast, cracks appeared – and a few determined women were quick to find their way in.

Hertha's claim to have been the first woman to read a paper before the Royal Society in person is more tenable. All papers put before the Society by non-members, whether male or female, were required to be 'communicated' by a fellow, with typically a third of papers received in this way each year. A Royal Society Yearbook spelt out the procedure explicitly: the communicating fellow was obliged to ascertain that the paper was 'fit and proper' whereon, according to the nature of the paper and other circumstances, the reading may consist of the title alone being read by one of the secretaries, the paper being read by the author or secretary

Table 7.1   Women publishing in the *Proceedings of the Royal Society of London* (A and B) and the *Philosophical Transactions of the Royal Society of London* (A and B) 1880–1914

	NAME	PROCEEDINGS Vol & Year	NO	PHIL.TRANS. Vol & Year	NO	TOTAL
1	Aston, Emily	56 (**1894**) with William Ramsey; 56 (**1894**) with Ramsey	2		0	2
2	Hertha, Ayrton	68 (**1901**); 74 (**1904/5**); 80 (**1908**); 84 (**1910**)	4	199 (**1902**)	1	5
3	Bate, Dorothea M.A.	71 (**1902/3**); 74 (**1904/5**)	2		0	2
4	Bayliss, Jessie S.	77 (**1905**) with Alfred J. Ewart	1		0	1
5	Beeton, Mary	65 (**1899**) with Karl Pearson	1		0	1
6	Benson, Margaret		0	199 (**1908**)	1	1
7	Boole, Lucy Everest	59 (**1895/6**) with Wyndham R. Dunstan and L.E. Boole	1		0	1
8	Brenchley, Winifred Elsie		0	204 (**1914**) with Hall and Marion Underwood (57)	1	1
9	Bruce, Lady (Mary)	85 (**1912**) (2); 86 (**1913**) (6); 87 (**1913**) (10); 88 (**1914**) (13) ALL with David Bruce, A.E. Hamerton and D.P. Watson	31		0	31
10	Buchanan, Florence	79 (**1907**)	1		0	1
11	Cave-Browne-Cave, F.E.	70 (**1902**) with Karl Pearson; 74 (**1904/5**)	2		0	2
12	Cayley, Dorothy M.	86 (**1913**)	1		0	1
13	Helen, Chambers	84 (**1911**) with S. Russ; 86 (**1913**) with Russ	2		0	2

Table 7.1 Women publishing in the *Proceedings of the Royal Society of London* (A and B) and the *Philosophical Transactions of the Royal Society of London* (A and B) 1880–1914 – *continued*

	NAME	PROCEEDINGS Vol & Year	NO	PHIL.TRANS. Vol & Year	NO	TOTAL
14	Chick, Harriette	71 (1902/3); 77 (1906)	2		0	2
15	Cuthbertson, Maude	83 (1910); 83 (1910); 83 (1910); 84 (1910); 85 (1911); 89 (1913) ALL with Clive Cuthbertson	6	213 (1914) with Clive Cuthbertson	1	7
16	Dale, Elizabeth	68 (1901); 68 (1901) with A.C. Seward; 76 (1905)	3	194 (1901) with A.C. Seward; 198 (1906) with Seward and Sibille O. Ford (24)	2	5
17	Dawson, Maria	64 (1898/9); 66 (1899/1900)	2	192 (1900); 193 (1900)	2	4
18	Drummond, Isabella M.	69 (1901/2)	1		0	1
19	Durham, Florence M.	74 (1904/5)	1		0	1
20	Embleton, Alice L.	78 (1906) with C. E. Walker	1		0	1
21	Fawcett, Cicely D.	62 (1897/8) with Karl Pearson; 68 (1901) with Pearson, Alice Lee (37), Warren and Agnes Fry (28)	2		0	2
22	Fawcett, P.G.	56 (1894)	1		0	1
23	FitzGerald, Mabel Purefoy	78 (1906); 82 (1910); 83 (1910); 88 (1914)	4	203 (1913)	1	5
24	Ford, Sibille O.	77 (1906) with A.C. Seward	1	198 (1906) with Seward and Elizabeth Dale (16)	1	2
25	Frankland, Grace C.	43 (1887/8) with P. Frankland; 47 (1889/90) with Frankland	2	178 (1887) with P. Frankland; 181 (1890) with Frankland	2	4

Table 7.1   Women publishing in the *Proceedings of the Royal Society of London*
(A and B) and the *Philosophical Transactions of the Royal Society of London*
(A and B) 1880–1914 – *continued*

	NAME	PROCEEDINGS Vol & Year	NO	PHIL.TRANS. Vol & Year	NO	TOTAL
26	Fraser, Helen C.I.	77 (**1906**) with Vernon H. Blackman	1		0	1
27	Fraser, Mary T.	81 (**1909**) with J.A. Gardner; 82 (**1910**) with Gardner	2		0	2
28	Fry, Agnes	68 (**1901**) with Pearson, Alice Lee (37), Warren and Cicely D. Fawcett (21)	1		0	1
29	Greenwood, Marion	54 (**1893**)	1		0	1
30	Huggins, Mrs. (Lady)	46 (**1889**); 46 (**1889**); 48 (**1890**); 48 (**1890**); 48 (**1890**); 49 (**1890/1**); 50 (**1891/2**); 51 (**1892**); 54 (**1893**); 61 (**1897**); 72 (**1903/4**); 72 (**1903/4**); 76 (**1905**); 77 (**1906**) ALL with William (Lord) Huggins	14		0	14
31	Isaac, Florence	88 (**1913**)	1	209 (**1909**) with Henry A. Miers	1	2
32	Johnson, Alice	37 (**1884**); 40 (**1886**) with Lilian Sheldon (54)	2		0	2
33	Kelley, Agnes M.	80 (**1908**) with F. W. Mott	1		0	1
34	Klaassen, Helen G.	54 (**1893**) with J. A. Ewing	1	184 (**1893**) with J.A. Ewing	1	2
35	Lake, Hilda	76 (**1905**)	1		0	1
36	Lane-Claypon, Janet E.	77 (**1905**)	1		0	1

Table 7.1 Women publishing in the *Proceedings of the Royal Society of London* (A and B) and the *Philosophical Transactions of the Royal Society of London* (A and B) 1880–1914 – *continued*

	NAME	PROCEEDINGS Vol & Year	NO	PHIL.TRANS. Vol & Year	NO	TOTAL
37	Lee, Alice	60 (**1896/7**) with Karl Pearson; 61 (**1897**) with Pearson; 61 (**1897**) with Pearson); 64 (**1898/9**) with Pearson and Moore; 64 (**1898/9**) with Pearson; 66 (**1899/1900**) with Pearson; 67 (**1900**) with Pearson; 68 (**1901**) with Pearson, Agnes Fry (28) and Cicely D. Fawcett (21); 71 (**1902/3**) with Pearson and Marie A. Lewenz (38)	9	190 (**1897**) with Pearson and Yule; 192 (**1899**) with Pearson and Moore; 193 (**1900**) with Pearson; 195 (**1900**) with Pearson; 196 (**1901**) with Pearson	5	14
38	Lewenz, Marie A.	71 (**1902/3**) with Alice Lee (37) and Pearson	1		0	1
39	Lloyd, Dorothy Jordan	87 (**1914**); 88 (**1914**)	2		0	2
40	Matthaei, Gabrielle L.C.	72 (**1903/4**); 76 (**1905**) with F. Blackman	2	197 (**1905**)	1	3
41	Maunder, Mrs A.S.D.	69 (**1901/2**)	1		0	1
42	Norris, Dorothy	84 (**1912**) with Arthur Harden; 85 (**1912**) with Harden	2		0	2
43	Nunn, Emily	34 (**1882/3**)	1		0	1
44	Ogilvie, Maria M.	59 (**1895/6**)	1	187 (**1896**)	1	2
45	Pixell, Helen L.M.	87 (**1913**)	1		0	1

Table 7.1   Women publishing in the *Proceedings of the Royal Society of London* (A and B) and the *Philosophical Transactions of the Royal Society of London* (A and B) 1880–1914 – *continued*

	NAME	PROCEEDINGS Vol & Year	NO	PHIL.TRANS. Vol & Year	NO	TOTAL
46	Porter, Annie	81 (**1909**) with H.B. Fanton	1		0	1
47	Raisin, Catherine A.	55 (**1894**); 55 (**1894**); 63 (**1898**)	3		0	3
48	Reid, Eleanor M.	82 (**1910**) with Clement Reid	1	201 (**1911**) with Clement Reid	1	2
49	Robertson, Muriel	81 (**1909**) with C. H. Martin; 85 (**1912**); 85 (**1912**); 85 (**1912**); 86 (**1912**)	5	202 (**1912**); 203 (**1913**)	2	7
50	Sargant, Ethel B.	65 (**1899**)	1	174 (**1883**) with Glazebrook and Dodds	1	2
51	Saunders, E.R.	62 (**1897/8**); 77 (**1906**) with Bateson and Punnett; 85 (**1912**)	2		0	2
52	Scott, Charlotte Angas	54 (**1893**)	1	185 (**1894**)	1	2
53	Seward, Margaret	45 (**1888/9**) with W.H. Pendlebury	1		0	1
54	Sheldon, Lilian	40 (**1886**) with Alice Johnson (32)	1		0	1
55	Sidgwick, Mrs. E.	34 (**1882/3**); 34 (**1882/3**); 37 (**1884**). ALL with Lord Rayleigh	3		0	3
56	Stopes, Marie C.	81 (**1909**) with K. Fujii; 85 (**1912**)	2	200 (**1909**) with D.M.S. Watson; 201 (**1911**) with K. Fujii; 203 (**1913**)	3	5
57	Underwood, Marion	71 (**1903/4**); 74 (**1904/5**)	2	204 (**1914**) with Hall and Elsie Brenchley (8)	1	3

Table 7.1 Women publishing in the *Proceedings of the Royal Society of London* (A and B) and the *Philosophical Transactions of the Royal Society of London* (A and B) 1880–1914 – *continued*

	NAME	PROCEEDINGS Vol & Year	NO	PHIL.TRANS. Vol & Year	NO	TOTAL
58	Waller, Alice M.	71 (**1902/3**) with A.D. Waller; 74 (**1904/5**)	2		0	2
59	White, Jean	81 (**1909**); 83 (**1910**)	2		0	2
60	Wilson, Marjorie	89 (**1913**) with H.A. Wilson	1		0	1
			148		30	178* (**173**) *Includes 5 papers counted more than once due to collaborations by women.

in part or in whole, or the author may be invited to give an oral exposition of contents, with experiments. On June 16 1904, there were fifteen other papers on the agenda for that afternoon in addition to Hertha's, including one by fellow Girton mathematician, Frances Cave-Brown-Cave. At the previous week's meeting a paper by geologist Dorothea Bate had been read and, later in the year, Florence Durham's paper on the skins of pigmented vertebrates was one of five papers read before the Society on December 1. In the absence of other evidence, it is difficult to ascertain whether any of these women, or indeed those which had gone before them, spoke in person. The Royal Society yearbooks, council minutes and journals do not state whether authors were present in person at the meeting in which their papers were read, so other corroboration is required. Hertha's presence can be verified by the details of John Perry's submission when he nominated her for the Hughes Medal and from Hertha's correspondence with Joseph Larmor, council secretary.[13] Referees had challenged Hertha's experimental findings and she had been allotted just ten minutes on June 16 to show an experiment to counter these objections.[14] Hertha did not therefore present her research findings in entirety as a paper before the fellows, she merely made a brief defence of her experimental procedures among an afternoon schedule of fifteen.

Hertha's paper was still not entirely accepted and just a short, revised abstract of it appeared in the *Proceedings* in March 1905.[15] The larger paper was not published in its entirety, in revised form, until 1910.[16] At some twenty plus pages long, reading the whole of this to the Society would have taken considerably longer than ten minutes. Hertha's papers were read before the Royal Society on six occasions: 1901 (on the mechanism of the electric arc); 1905, 1908, 1911, 1915 (on the formation of sand ripples); and 1919 (on a new method of driving off poisonous gases). By 1914 women were permitted as a matter of course to attend the meeting at which their paper was read, so long as they were accompanied by a fellow.

The celebration of Hertha as an exceptional female presence at Royal Society conversaziones also requires contextualisation. The Society held two conversaziones at their premises in Burlington House each year; the first in May was colloquially named 'the black one' as, explained *The Times* on May 10 1900, 'it was exclusively confined to the sombre sex' with exhibits that 'would appeal to the specialist'. The second conversazione of the year, held in June, was generally known as 'Ladies Night' and reported in the press as a society function. Characterised as 'less severely scientific',[17] this offered refreshments as well as scientific entertainment, with tea and coffee served in the Officers' Rooms and wines and ices on the ground floor. It was at these 'Ladies Soirees' that Hertha participated, her first appearance being on June 21, 1899, when she displayed her experimental apparatus in the library alongside twenty-two other exhibits. Although she was the only female exhibitor on this occasion, the following year astronomer Annie Maunder displayed her photographs of the Milky Way and, in 1904, Hertha's illustration of the formation of sand ripples was joined by scientific filmmaker Mrs. D.H. Scott's display of new techniques in photography showing the movements of plants.[18] Women then, although rare, were not entirely absent at the periphery of the Royal Society. Similarly, women's names can be found as recipients of Royal Society grants – Minutes of Council show awards being made to Dorothea Bate,[19] Gertrude Elles[20] and Edith Saunders.[21] It seems that the vexed relationship of women with the Royal Society was fluid and changing at the turn of the nineteenth century. The interesting question to explore is why the Society dallied with women at its margins yet was wary of accepting them in any formal, explicit way.[22]

Politics in the fellowship

There seems evidence at turn-of-century that senior scientific women felt their exclusion from the Royal Society acutely and were organising with determination to redress the situation. The first formal approach lobbying for the admission of women fellows was in 1900. Marion Farquharson,[23] a cryptogamic botanist (specialist in the evolution of plants), fronted the request with support from Dr. Elizabeth Garrett Anderson and members of the Agricultural Association for Women. Farquharson believed, as did Hertha, that being denied a fellowship directly impacted on her ability to pursue research, a situation compounded by her exclusion from both the Linnaean Society and the Herbarium of Linnaeus. In reply to Farquharson's request, the Royal Society stated that the question of women's admission was determined by its Royal Charters which had been interpreted as extending only to men.[24] The 'interpretation' of charters was also to be an issue two years later when, again, the question of admitting female fellows raised its head.

When John Perry proposed Hertha Ayrton for a Royal Society fellowship in 1902, the receipt of her certificate of candidature presented the council with a dilemma that touched on more than just the Society's relationship to women. Hertha's nomination also served to foreground questions about the Society's position in relation to other learned societies and raised issues surrounding its standing and status both within science and the wider governing class. For the Royal Society, the key mechanism that gave self-identity, status and prestige was the selection of its fellows. Who should be eligible for nomination and how their election should take place had occupied the Society from almost its earliest days and, during the nineteenth century, moves had been made to limit membership and enhance exclusivity. The role of 'amateurs' and 'dilettantes' was successfully curtailed in 1847 by a revision of statutes which included restrictions on eligibility, number of nominees and election procedure. In previous decades just a third of fellows had been scientists, the rest using the Society as an exclusive club. Now only proven scientists could be nominated and they needed the support of six, instead of three, fellows. A limit on numbers was also introduced, a change which ensured that only one election took place each year and that no more than fifteen new fellows were elected. The power of the president and council was enhanced significantly as the task of recommending the names to be put forward for election was placed in their hands. Not surprisingly, the number of men vying for election greatly outnumbered the fellowships available and this was a factor in the comparatively slow infiltration of 'commercial men' into the Royal Society. It was not until the second decade of the twentieth century that professional scientists comfortably outnumbered the gentleman amateurs thanks to the development of electricity and other new technologies.[25] At the time of Hertha's nomination, involvement in commercial enterprise could, by raising the question of self-interest, still taint a researcher's work and imply conflict with the aims of the Society. As William Huggins reiterated in his 1903 presidential address, the unique position of the Royal Society among other academies had been reached '... by its unwearied pursuit of truth for truth's sake without fee or reward'.[26]

Huggins's implied criticism of fellows who worked not out of ideals of service but for profit, reflected a fault line within the Royal Society and could have been aimed directly at the 'modernisers' who had put forward Hertha's name the previous year. Of the fellows who signed her nomination, John Perry worked with William Ayrton providing technical education and developing commercial applications at the Central Institute. The other signatories were William Preece, who was chief engineer at the Post Office and had introduced wireless telegraphy to that organisation; William Tilden, chemistry professor at the Royal College of Science, a partner institution of the Central at South Kensington; Raphael Meldola, professor of chemistry at Finsbury Technical College and member of the Society of Chemical Industry; William Abney, an astronomer and photographic chemist who served on the Board of Education and was a member of the Society of Chemical Industry; George Carey Foster, a physicist/chemist who had just been appointed Principal of University College London; Olaus Henrici, a physicist who lectured at the Central for the mathematics and mechanics department and held a chair at Bedford College London; Joseph Everett, physicist, and Norman Lockyer, astronomer and editor of *Nature*. Both Lockyer and Carey Foster were noted for their sympathy with women's causes. Lockyer supported the admission of women to learned societies in his journal; Carey Foster had been one of the pioneers of the 'London Ladies' Educational Association' which had led to the admission of women to London University in 1878.[27]

Royal Society President William Huggins was bitterly opposed to Hertha being elected a fellow; it was during his presidency too that she was nominated and, in 1906, awarded the Hughes Medal. On the latter occasion Huggins wrote to Joseph Larmor, Secretary to Council, blaming his cold for preventing him from taking the Chair that day and implying that, had he been present, he would have obstructed award of the medal.[28] Huggins had co-authored several astronomical papers with his wife Margaret, but their collaboration was not presented to the world as an equal partnership. As noted in Chapter 4, Barbara Becker has demonstrated how the couple colluded to present a traditional and romanticised image of themselves with William as the principal investigator and Margaret his loyal and gifted assistant.[29]

Huggins's antipathy to scientifically-presumptuous women, and to commercialism, found a joint target in the controversial person of Hertha Ayrton. Hertha's research on the electric arc had led to patents for applications to streetlights, searchlights and cinematography, so she too was a worker at science's technical and commercial fringe. For a male scientist, this raised questions regarding the objectivity of his work. For a woman, it also transgressed ideals of feminine service and defied the assumption that respectable, middle-class women did not receive payment for their work. The relationship between science and technology was hierarchical yet fluid at the end of the nineteenth century and one of the functions of the Royal Society, achieved via the selection of fellows and acceptance of papers, was to police and maintain this divide. Huggins and Joseph Larmor, who regarded William Thomson/Lord Kelvin's commercial interests as distractions from the proper intellectual pursuits of a natural philosopher,[30] represented the elitist ideals of an old guard for whom the taint of commercialism and the challenge of the professional men (and possibly women) represented a threat to the status of the Royal Society, the status of science – and the status quo.

A similar clash of class and ideology was being played out at Cambridge, another institution facing an influx of the middle classes and adaptation to meritocracy. Here, the infiltration of rational, secular, scientific ideas was challenged by dons who, by promoting the late-Victorian concept of the 'ether', championed non-material interests and the 'imagination' as opposed to the 'cowardly security' of sense data.[31] In line with romantic ideology, such an approach privileged 'innate ability' over 'acquired skill' and served to set limits as to who was eligible to participate in scientific discovery. (It is no surprise then that pure mathematicians Grace and William Young were scathing in their criticism of materialist scientific theory, arguing, for example, that the views of Thomas Huxley and John Tyndall were 'crude' for holding 'that everything could be explained by means of mechanical, chemical, electrical etc. laws'.[32]) The struggle being played out at the Central to achieve status for technical education and engineering (described in Chapter 3) was brought by its protagonists to the Royal Society. These pioneers of technical training for the professional middle classes were attempting to repudiate the notion that practical labour was inferior to intellectual labour, and to democratise science by opening it up to more than just a 'natural born' elite. In this light, Hertha's proposed fellowship can be interpreted as part of a larger battle for the soul of science; her nomination another tool in the moderniser's armoury of strategies to promote change.

The culture of the Royal Society

Just as sections of the Cambridge elite clung to metaphysical concerns in the struggle against change, so the Royal Society carved a self-identity and cohesiveness based on a history, language and rituals that were all exclusively male. Creating a continuity between existing fellows and the natural philosophers who had established the Royal Society in the seventeenth century was key to this process. At each meeting of the Society and council, the mace, presented by Charles II, was still placed on a table in front of the president; in addition, all new fellows were required, as their earliest predecessors had been, to sign the charter book (this practice is still current today). In 1857 the Society had moved to Burlington House in Piccadilly – a palatial mansion (now housing the Royal Academy of Arts) which, like the Royal Society itself, had its origins in the 1660s. In 1873, transferral to the specially-refurbished East Wing had provided ample space to display the many portraits, busts, statues, medallions and other relics that were in the possession of the Society. These reminded fellows of their heritage and reinforced their consciousness of being one of a highly-select group of men. The only representation of a woman was a bust of Mary Somerville who, with Caroline Herschel, was made an honorary fellow in 1835. It was one hundred years later before a woman was admitted in person to such a privilege; the first female fellows were not elected until 1945.[33]

When *World Magazine* profiled Sir Michael Foster, secretary to the Royal Society, at his rooms in Burlington House, special reference was made to him being 'surrounded by objects associated with the great pioneers of science'.[34] Amongst portraits and statuettes of Darwin, Cook, Newton and Humboldt, there was a vast oil painting that took pride of place above the fireplace. This depicted the president and members of the council of the Royal Society waiting upon Michael Faraday, asking him to accept the office of president. This image of scientific 'apostolic succession' – a religiosity recalling classical paintings of Christ and his disciples – was enhanced by a number of relics and instruments retained by the Society. These included a lock of Newton's hair, two rules purportedly made from the wood of his apple tree, and a cast of his face taken after death. Such imagery also hinted at a male asceticism, long associated with intellectuality, which was important at a time when rationalism was being challenged by the rise of movements such as decadism and spiritualism.[35]

Maintaining ritual reminiscent of the Christian church may also, for some fellows, have been a way of maintaining a comforting continuity and structure after their religious faith had been successfully challenged by Darwin. Recalling the male-centred practices of the clergy, it certainly presented another obstacle to women being accepted in the similarly all-male community of the Royal Society.

Two other institutions to which many Royal Society fellows belonged reinforced this male exclusivity: the Athenaeum Club and the Royal Society dining club. Upper-middle-class males could be uncomfortable around women, their lives being structured around institutions (boarding school, Oxbridge colleges, the Regiment) where women were rarely present except in roles such as maids, matrons or cleaners.[36] In 1911 such unease was articulated in a book arguing that, in the interests of courtesy, gentlemen cannot tell ladies that they are wrong, therefore meaningful intellectual discussion between the sexes is impossible.[37] The difficulties of frank exchange between men and women was also a theme of the literature of the time, especially in novels by 'New Woman' writers such as Sarah Grand and Olive Shreiner. As one of Grand's cross-dressing characters in *The Heavenly Twins* confides, real understanding between the sexes could be hard to achieve in a sexually-segregated society: 'I have enjoyed the benefit of free intercourse with your masculine mind undiluted by your masculine prejudices and proclivities with regard to my sex'.[38] Earlier similar sentiments had been ridiculed by Edwin Abbott in his satire *Flatland*: 'among Women, we use language implying the utmost deference for their Sex ... but behind their backs they are both regarded and spoken of as being little better than mindless organisms'. Abbott also alluded to the 'double-training' given to children with boys being removed from their mothers and nurses at the age of three to be taught a new 'vocabulary and idiom of science'.[39] Abbott may have had his tongue firmly in his cheek, but Almroth Wright, FRS, was wholly serious when he suggested in *The Times* in 1912 that men can only do their best work when free from 'the onus that all differential treatment imposes'.[40] Situated just around the corner from the Royal Society and long associated with it, the Athenaeum's namesake was the ancient Roman centre for the study of literature and science. Many fellows (including Grace's husband) patronised this gentlemen's club which offered a seamless passage for men of science from one comfortable and select all-male environment to another.

Many of the elite of the Royal Society also held ambitions to join the Society's dining club. Unlike the president and secretaries, fellows had no automatic right to membership but had to be proposed in writing

by three 'diners' and then elected, with at least a three-quarters majority, by the some sixty or more members. The rules of the club give an
insight into how a late-nineteenth-century male network of patronage
and power worked in practice. Each ordinary member had the privilege
of inviting one guest to dine each year and visitors included foreign men
of science, civil servants, MPs and members of the government. It was a
mechanism by which the Royal Society could extend its influence and
gain status over other learned bodies. For example, fellows were often
seconded to government commissions, therefore the letters 'FRS' after
your name could be a gateway to a government-funded management
position. In 1902 the new National Physical Laboratory was put under
Royal Society control and Richard Glazebrook, FRS, was appointed director. In the same year Glazebrook was also elected to membership of the
Royal Society dining club.

This segregation of the sexes, which so dominated the habits of
the Royal Society, was predicated on a gendered conception of science
that conceptually, and effectively, removed any notion of femininity
from the public practice of science. The sexual metaphors of Baconian
science, the construction of woman as object/nature, and man as the
virile subject/penetrator of her secrets, are familiar and often rehearsed.[41]
This representation of women as the object of enquiry (but never the
active enquirer) infused the ideals of the seventeenth-century founders of
the new natural philosophy – ideals of which fellows in the nineteenth
century were constantly reminded by the iconography and language
of their surroundings. It is also important to remember that the first
fellows explicitly sought to create an unambiguously separatist, 'masculine science', based on a new materialist vision, in order to differentiate
themselves from alchemy, a practice which still provided an alternate and
competing model of nature. In the latter tradition, matter was believed
to be suffused with spirit and both male and female principles were necessary for understanding. While the central image for the new mechanical philosophers was man subduing and controlling a female nature,
the alchemists made metaphorical use of images of coition, the merging
of male and female and the conjunction of mind and matter. Paracelus
wrote that a man without a women is not whole – for him the sexes were
(allegorically if not in reality) equal.[42] The commitment to a 'masculine
philosophy' heralded a start to femininity being regarded as not different, but inferior, a trope reiterated in the Darwinian theories of the late
nineteenth century.

In a preface to his 1903 Presidential Address, William Huggins
emphasised the Royal Society's status in relation to the new societies

by characterising it as 'The Mother Society' and the newer bodies as her 'daughters'.[43] Such characterisation was not atypical; it was used again when the Royal Society was praised as 'the Mother and Model of all the learned societies in the English speaking world'.[44] Given the Royal Society's early emphasis on masculine science, feminine iconography did not flourish in English science,[45] so this later representation of 'mother' and 'daughter' societies is interesting. It has parallels with the 'mother' Church – a 'mother' society with its own priesthood of ascetic men of science and its own secular rituals such as those outlined above. This discourse was similar to that prevalent in freemasonry. A newly established lodge was the 'daughter' of an older 'mother lodge' and the male lodge members identified themselves as 'brothers'.[46] This type of language and metaphor can also be interpreted in the context of an Edwardian glorification of the role of motherhood amidst controversy over sex roles, a declining birth rate and the influence of the eugenics movement. Either way, the implication is that a woman's role is to give birth: to facilitate the acquisition of knowledge, not to produce it herself. So, at the Tercentenary celebrations of the Royal Society in 1912, while the Dean of Westminster Abbey assured fellows and international delegates that 'through the pre-eminent influence of the men of the Royal Society ... Reason, as the noblest gift of God to man ... [will continue] ... the passionate search for the secrets of truth',[47] women were preoccupied with more prosaic tasks. A 'Committee of Ladies' was formed to provide entertainment for wives accompanying delegates. The meeting room at the Royal Society was put aside for their use, and here they waited to be of service, wearing different coloured badges to indicate which languages they spoke.

The 'irrational' face of femininity

If these women offered a comforting image of womanhood to many fellows of the Society, the similarly dressed, predominantly middle-class suffragettes (who were engaged in increasingly militant agitation for the vote around 1912) must have inspired exactly the opposite. Some of the most disturbing episodes of the suffrage campaign took place literally around the corner from the Royal Society, including fierce confrontation with the Police at Buckingham Palace, suffragettes chaining themselves to railings and 'rushes' on Parliament. On November 18 1910 Hertha was involved in 'Black Friday' when protesters, marching on the Commons, met with violence and brutality from the Police. In 1912 a campaign of mass window breaking began in the West End, along with post boxes being set alight and the cutting of telephone/telegraph lines. In May 1914, after Mary Richardson slashed the Rokeby Venus, public galleries and museums (including those at South Kensington, frequented by many fellows) were temporarily closed.[48] Hertha joined Mrs. Pankhurst's militant WSPU soon after its establishment in 1903. Although she never engaged in law-breaking, Hertha did help co-ordinate opposition to the 1912 census boycott and, very publicly, nursed Mrs. Pankhurst at her home when the WSPU leader was released and rearrested a number of times during 1913 under the Prisoners' Temporary Discharge Act (the Cat and Mouse Act). As a well-known scientist, Hertha was also asked to lobby scientific men to sign a petition in favour of female suffrage, a task that she did not enjoy. Amongst the notable refusals were Francis Galton FRS, father of eugenics, and Sir Archibald Geikie FRS, President of the Royal Society.

It is clear that suffrage campaigners challenged what they saw as the male homogeneity of science and that women's access to science became one of a number of pivotal issues around which protest was articulated. Suffrage banners were designed around the names of notable female scientists, including 'Marie Curie, Radium', 'Caroline Herschel' and 'Mary Somerville', and the display of these on a protest march conveyed strong messages about female intellectual equality in an important 'male' sphere.[49] Within the context of this co-ordinated assault on the male privileges of science, Hertha's commitment to the suffrage campaign was enough to cause a mixture of condescension and anxiety in then Royal Society President William Huggins. When Hertha was awarded a Medal in 1906, his letter to Joseph Larmor supposed that there would be 'great joy and rejoicing in H.M.'s gaol, among the women in prison' and that Girton and Newnham Colleges would 'get up a night of orgies on the 30th in honour of the event!'. This may well have been a reference to the highly-publicised celebrations that took place well into the night around a bonfire at Newnham College on the occasion of Philippa Fawcett's success at beating the senior wrangler in 1890.[50] But the tone of amusement is absent when he asks 'Can we now refuse the Fellowship to a Medallist?'.[51]

Huggins' image of sexually out-of-control women engaging in 'orgies' was in line with arguments made by opponents of women's suffrage and higher education who pointed to women's emotional motivations and questioned female rationality. Almroth Wright's 1912 letter to *The Times* went on to warn about 'militant hysteria' and to argue that women who assert intellectual equality with men are 'plainly' displaying 'an element of mental disorder'. Sir Edward Almroth Wright was Professor of Experimental Pathology at London University and a respected member of the scientific community; he had been a fellow of the Royal Society since 1906 and had exhibited alongside Hertha at the ladies' conversazione in 1904. His opinions on the shortcomings of the female intellect were shared by earlier fellows. Francis Galton FRS, member of the anti-suffrage society and Athenaeum Club, had used his research into heredity to 'prove' that women were defective in muscular power and sensory and intellectual activity; his investigation into the abilities of Royal Society fellows, *Englishmen of Science: Their Nature and Nurture*, had been published in 1874. Similar views articulated by Central chemist Henry Armstrong, a one-time member of the Royal Society council, have been outlined in a previous chapter. Armstrong was vocal in his opposition to the admission of women to the Royal Society, his objections centring on the incapacity of women to be trusted with impassionate, objective scientific method.

Recent scholarship within the history of science has explored the ways in which a piece of work becomes accepted as knowledge. Shapin and Schaffer argue convincingly that this is a social process that depends on the existence of a scientific community that shares a set of social codes and conventions, and within which each (gentleman) member is accorded trust and respect as an equal and a peer.[52] Gender has seldom been addressed within the context of these arguments, yet issues surrounding the access of women to science (and to the Royal Society in particular) at the end of the nineteenth century touch directly on concepts of trust and perception of trust. (These issues have been discussed with reference to women and the laboratory in Chapter 5.) Darwin had theorised that women's intellect was not on a par with man's and that women were lower down the evolutionary scale, closer to the animals. Women operated on emotion and instinct but men, especially men of science, had the natural gift of rationality. For much of Darwinian-influenced science, women's reasoning was untrustworthy, therefore women's conclusions were questionable. Women could not 'by nature' have equal status as scientists, or justifiable ambitions to join an elite scientific society. As the 'hysterical' suffrage campaign gained momentum, with increased militancy, hunger strikes, and the force-feeding of women at Holloway gaol, this view gathered urgency and was used as another layer of evidence to support the anti-woman case.

Hertha and the Royal Society: Discord and difficulty

Although it is difficult to be certain about the motivations of referees, it is possible to point to a growing tendency to question Hertha's scientific reasoning during the first decade of the twentieth century. As early as

1901, a Royal Society referee for her paper on 'The Mechanism of the Electric Arc', which had been so well-received by the Institution of Electrical Engineers (IEE), challenged her understanding of key scientific concepts and commented on her 'faulty or at least doubtful reasoning'.[53] Despite this, the paper was accepted for publication in the *Philosophical Transactions*. Her 1904 paper, 'The Origin and Growth of Ripple Mark', was less well received and resulted in a dispute that was to last for the next eleven years. Referees Horace Lamb and John Joly[54] both took issue, advising against publication in the *Transactions* (the outlet for longer papers offering complete research) and offering only reserved support for its appearance in the *Proceedings* (which had originated as a vehicle for abstracts and Society news, and which now carried shorter papers and research in-progress). Joly criticised Hertha's proof as 'crude' and accused her of forming 'wide conclusions on apparent slender evidence'. Lamb considered that since 'the dynamics of much simpler phenomena ... is only very improperly understood, an exact theory cannot be expected' and can 'only inspire a qualified and provisional confidence'.[55]

Lord Rayleigh and Professors George Darwin and Osborn Reynolds had all failed to explain the phenomena fully, so there was an implication that it was over-ambitious for a woman to have claimed to have done so. These male scientists had position, titles and standing, the latter two attributes reproduced at the top of their papers alongside their names. These 'endorsements' of their veracity and, therefore, the trustworthiness of their work, can be interpreted as a later version of the gentlemanly codes that Shapin and Schaffer identified as crucial to the production of scientific knowledge. Hertha had none of these and her work was not taken on trust. Because of these 'unsatisfactory' theoretical explanations, at the June conversazione her planned oral exposition was replaced by an experimental demonstration. Despite further elaborations read in 1908 and 1911, the Royal Society refused to publish her paper unless she removed the theoretical explanations and deductions. The issue was not fully resolved until 1915 when a paper answering the criticisms, communicated (and therefore endorsed) by Lord Rayleigh, was accepted.[56] Had Hertha's suffrage work prejudiced her position? She certainly thought so, as she expressed in dismay to friend and fellow electrical engineer A.P. Trotter.[57]

Hertha was in her fiftieth year when her 1904 paper was questioned and sixty-one when it was finally accepted and published by the Royal Society – an elite body which accepted as fellows only men of experience and standing who had made a marked contribution to science. In this context, maturity was a benefit to men, but for women it could present difficulties. Just as young women, for example the students at the new women's colleges outlined in Chapter 1, were typically represented as flowers, so older, menopausal women could be described using the metaphors of biological decay, their reproductive organs 'withering on the vine'. Importantly, as a woman's bodily health and mental well-being was linked to her reproductive biology, the onset of menopause was not just associated with physical decay but with moral and intellectual degeneracy too. This representation was common in the medical press and advice manuals, both of which exhibited an increasing interest in the menopause in the context of contemporary concerns about motherhood. As late as 1915, one specialist journal asked its male readers to picture their mother-in-law's 'want of sweet reasonableness and her lack of charm' and to consider how her womanliness 'disappears to be replaced by the moustache, the bearing and the assertiveness of the male'.[58] Again, it fell to Almroth Wright, FRS, to stress that even most women acknowledged that 'half the women in London need to be shut up when they come to the change of life'. He warned readers and doctors never to 'lose sight of the fact that the mind of woman is always threatened with the reverberations of her physiological emergencies'.[59]

Hertha faced numerous difficulties in her later career in addition to the disagreement with the Royal Society over her disputed theoretical conclusions. This latter dispute developed into a long and bad-tempered correspondence. Hertha wrote requesting a copy of the referees' reports, a request which the Society was very reluctant to comply with, fearing controversy. After three more letters from Hertha objecting to this decision, the Royal Society secretary finally gave way and, in a terse note, replied that he would provide her with a copy, 'under pressure' and 'for finality's sake'.[60] Later, Hertha found it almost impossible to fight official apathy and have her design for an anti-gas fan at least appraised. At the outbreak of World War One, Hertha decided to apply her research on water vortices to the air and designed a fan that could create 'eddies' or vortices of pure air that would drive poison gas away. Trotter describes her experiments in her home laboratory:

> To imitate a gas cloud she used the smoke of brown paper, but this while warm tended to rise above her laboratory battle-field ... (so) ... cooling chambers and pipes were devised and made, and smoke poured out and rolled along the floor. A few flaps with a card on a matchbox serving as the parapet of a trench drove it back.[61]

The simple device that ensued, a paddle-shaped fan or 'blade', mounted on a T-shaped handle, which was about a third of a metre square, did not immediately convince the authorities.[62] At the Central Gas Laboratory developments were being made of a far more complex nature, involving large-scale and 'handy' fans with carbon filters and motors, supervised by Royal Society fellows, Professors Watson and Haldane.[63] Again, Hertha's absence from material and social networks of credibility and trust hampered her ideas from receiving consideration, let alone acceptance.

This marginalisation may have been compounded by the loss of her husband, William Ayrton, who had died six years previously. Ayrton had been a fellow of the Royal Society, an influential figure in the London scientific community and a strong, vocal supporter of Hertha's work, both privately and in public. Hertha could not persuade War Office officials to take her seriously and travel to her home to evaluate these 'toy-like models in her drawing-room'.[64] If her research had been carried out at a prestigious institutional facility, or if she had been able to add the endorsement of 'FRS' after her name, they may have responded with more alacrity and assessed her work without fearing any threat to their own credibility. In the end, a large number of fans were sent to the Front but their efficacy was a subject of dispute, not least in angry exchanges between Hertha and her critics on *The Times* letters page.[65] At least one account of that period implies that she was the tiresome 'wife of a distinguished physicist' (although William Ayrton had died in 1908) whose fan was only accepted after officials caved in, exhausted, to pressure. The writer added that the best use of the fan had been to burn its wooden handle for emergency fuel.[66]

Distinguished men – redundant women: Differing significations of age

Appropriate occupations for older women who were past child-bearing age was a discussion largely absent from the controversies over women's role in the decades around 1900. That very year one contributor to the medical journal *The Lancet*, imagining no useful occupation for women past reproductive age, suggested that they should accept 'voluntary elimination' for the good of the race.[67] Opponents of the suffragettes depicted them as old and ugly spinsters and witches; the image of the 'Dark Widow' was constantly employed, referring in particular to suffragist Mrs. Fawcett and militant leader Mrs. Pankhurst. This image was also suggestive of Jewishness: dark and 'crone-like', not unlike the sinister image of the 'medical flapper' discussed in Chapter 3. It could be that Hertha's age, her Jewish background and stereotypical 'dark' hair and looks, together with her public association with suffrage, may have influenced detractors (perhaps unconsciously) against her. Far from claiming a worthwhile role for older women, the suffrage movement's response to such representation was to relocate their leaders back into the accepted norms of femininity and, in their propaganda and imagery, to stress youth. Mrs. Fawcett was represented in the literature of the National Union of Women's Suffrage Societies (NUWSS) as above all a wife and mother; the militant WSPU retaliated with ubiquitous images of Joan of Arc – young, beautiful and noble.[68] A reluctance to foreground older women and the difficulty of showing positive representations of them is illustrated by the fictionalised biography of Hertha, *The Call* (1924), written by her step-daughter Edith Ayrton Zangwill. Here episodes from Hertha's life are transposed onto 'Ursula' who is a much younger, beautiful woman.

For the Royal Society there was little nervousness at representing men of a certain age in a positive way: images of mature men of science, in the form of portraits, busts and medals, lined the walls of Burlington House. The election of only older, experienced men who could be proved to have made a significant contribution to science, was one of the key ways in which the Royal Society maintained its difference from other learned bodies. As William Huggins stressed, the admission of young, less experienced men would necessarily take from the Society 'its select and exclusive character, and its distinctive position as an Academy'.[69]

The Royal Society and younger scientific associations: Maintaining the hierarchy of science

The relationship between the Royal Society and other predominantly younger, specialist societies, was a matter of concern amongst fellows at turn-of-century. In 1903, Huggins devoted his Presidential Address to the 'grave question' of how long the Royal Society could maintain its 'high position of distinction and influence' without reforming its co-ordination with the specialist societies. The establishment of new learned bodies had been a fast-growing phenomenon in the last quarter of the nineteenth century and between 1860–1900 journals of professional societies in Britain increased from around twelve to seventy.[70] The other broadly-based scientific society was the British Association for the Advancement of Science (BAAS). This had been established in 1831 with the aim of communicating science to the public and building bridges between science and industry. Meetings were held around the country, at industrial and university towns, where series of popular lectures designed to attract a broad, amateur audience were delivered. Attendance at meetings did not require a scientific qualification and, after early opposition, women were admitted as members in 1848. However there was controversy well into the twentieth century as to whether women should be eligible to serve as officers and it was not until 1913 that the first woman, botanist Ethel Sargant, was elected as section president.[71] Hertha presented four papers at BAAS meetings in the 1890s and 1900s, although she was never registered as a member. Unlike the Royal Society, BAAS meetings were recognised as social gatherings as well as scientific ones with special 'ladies tickets' available until 1919. One of the ways that the Royal Society marked its elite, scientific superiority over the BAAS was to retain an all-male membership. The Royal Society was not alone in using women (or their absence) as signifiers of status and seriousness. Darwinist Thomas Huxley had prevented women's admission to the Geological Society and engineered their exclusion from the Ethnological Society specifically to upgrade its professional status in relation to the breakaway anthropologists.[72]

One of the dynamics driving the birth of new societies was the need of amateurs, including women, to find an outlet for their enthusiasm and work when these were refused by older, elite societies. For example, the British Astronomical Association was founded in 1890 to provide an alternative to the Royal Astronomical Society (established 1820) and was advertised as 'open to Ladies as well as Gentlemen'. Several women were active in the Association, participating in expeditions, serving on its council and editing its journal.[73] The Physical Society too, before which Hertha presented two papers,[74] had been established in 1873 partly to provide an outlet for incomplete work which would not have been accepted for publication by the Royal Society. Membership consisted of university and school teachers of physics and amateurs; women had been eligible for membership from the beginning.[75] For many of the newer societies that did not possess a royal charter, the concern was not to exclude members, but to attract them and forge a particular identity. The IEE had been established in 1871 as the London-based Society of Telegraph Engineers. When Hertha lectured before them in 1899 the event attracted welcome attention from the press; IEE members worked at the intersection of commerce and science and were in the vanguard of developing technical applications for public and private use. Publicity that caught the public imagination, of male and, especially, female 'consumers' who may have a say over the use of electrical appliances in the home, was useful to their image and to their business. After speaking before the IEE, Hertha gave a lecture on the electric arc at the populist Electrical Exhibition at Olympia.

Hertha was elected a member of the IEE at the same time as her lecture but her admission papers show that she was admitted under an exceptional clause and did not follow the same election process as the average male. This special clause did not require any electrical education or employment qualification, merely that the candidate 'shall be so prominently associated with the objects of the Institution that the Council considers his admission to Membership would conduce to its interests'.[76] It was not until 1919 that the next woman became connected to the IEE when Gertrude Entwisle was elected formally as a graduate member. Entwisle was of a retiring nature, worked away from London (in the North West) and did not interact with the press as Hertha had done before her. As a result, the attendance of this female engineer at IEE gatherings was sometimes unexpected: 'When she attended her first IEE meeting, the lecturer mistook her for a militant suffragette and stopped all proceedings' and after her election 'it took half an hour and the special pleading of the Secretary of the North Western Branch to gain admittance [for Entwisle] to the Manchester Engineers Club to attend an IEE meeting'.[77]

Hertha's nomination to the Royal Society

In the context outlined above, the exclusion or inclusion of women was an important tool in defining a society's identity. When they nominated Hertha in 1902, the 'modernisers' were aware that similar campaigns on behalf of women were being rehearsed at other august institutions. The Linnaean Society and the Royal Astronomical Society, both of which shared premises with the Royal Society at Burlington House, had debated the issue. The latter reiterated its refusal to admit women in 1892 when three women were put forward as candidates. The Linnaean Society finally gave way and opened its doors to women in 1904. The Entomological Society, 'formally so exclusive that ladies who contributed papers were not even admitted to be present when they were read'[78] elected its first woman member in 1904 too. Disputes over the admission of women to learned societies had a habit of breaking out onto the pages of the national press instead of remaining behind closed doors. When the Royal Geographic Society debated the possibility of female fellows in 1892–3 an angry dispute between council members was conducted via the letters page of *The Times*.[79] The Royal Society, in keeping with its attachment to an austere, exclusively-male past, turned Hertha down. But rather than cause controversy and dissent among fellows by a straightforward refusal, it argued that its Charter forbade a certificate of candidature from a married woman to be registered or read. As a married woman, Hertha's status in law was covered by that of her husband, therefore she could not be elected a fellow. The council had sought a legal opinion and this, after much contradictory argument, had advised against the eligibility of married women. The lawyers were less sure about the position of unmarried women, prevaricated, and left the issue in the hands of the council (who ignored it). To admit unmarried women would have required a change to the statutes, something that had been carried out more than once in an effort to control the nature of the membership and election procedures. If the Royal Society had possessed the will to admit married women it could have applied for a supplemental charter as the Royal Astronomical Society had done in 1915 when it finally accepted women.[80]

To president Huggins and councillors Larmor and Armstrong, all of whose negative views on women were well known, allowing discussion of Hertha's nomination may have been interpreted as too dangerous, threatening dissent amongst fellows and criticism in the press, not least in Norman Lockyer's journal *Nature*. (Huggins and Lockyer, astronomers both, were acknowledged as 'arch rivals' by this time and Hertha's election was intimately connected with personal animosities within the Society.) What's more, the Royal Society may have been forced to 'capitulate' as some members of the council were known sympathisers with women's issues. Naturalist William Bateson was a vocal supporter of degrees for women and had founded a 'school' of genetics at Cambridge University comprising primarily of women from Newnham and Girton Colleges; Michael Foster was supportive of higher education for women and allowed female students to attend lectures in his physiology laboratory at Cambridge; astronomer H.H. Turner was a friend of John Perry's and held similar, pro-women opinions. The nervousness of the council over this issue, and their desire to avoid dissent, is evidenced by the letter, sent to every fellow, explaining that legal opinion had instructed that Hertha's nomination could not be registered or read. In this way the council ducked responsibility for the issue by implying that their hands were (legally) 'tied'. The refusal of a fellowship directly affected Hertha's scientific research and to a certain extent isolated her from developments in her field (as Marion Farquharson had complained in 1900). In 1910 Hertha was moved to ask the then president, Sir Archibald Geikie, if she could have the privileges of a fellow in the one respect of receiving unpublished papers, as not having this access, she wrote, had put her at 'a great disadvantage'.[81] No evidence can be found of Geikie's response. However the relation of women to the Royal Society became a festering sore which refused to go away.

Women's complaints go public

On June 16 1914, on a day coinciding with the Royal Society annual soirée, an anonymous correspondent mounted a long, passionate and blistering 'Complaint against the Royal Society, The Handicap of Sex' which was published in full in *The Times*.[82] Much of the argument centred on the fact that on the one night when the Society's doors were opened to the public and women 'mingle with the hoary-headed scientists', it was the *wives* of scientists and not *woman scientists* who were granted admittance:

> But any Amelia or Leonora whom chance married to a scientific man is eligible as his 'lady'. Women high up in scientific positions, women with international reputations, women who would themselves bear the magic title of F.R.S. if they could disguise from the world the fact of their sex – such women are shut out from the concourse of their intellectual fellows, shut out from the opportunities of meeting and talking with their scientific colleagues, unless they know by chance some bachelor Fellow, or one whose wife does not care to show off her diamonds, who will take her *incognito* as his 'lady'.

The article makes criticism of almost all of the Royal Society's practices, including admission to conversaziones, biased publication policy and the requirement of distinguished women to find a 'communicator' for their papers. It concludes with a call to 'let the world never forget that every discovery published by a woman represents much more than the same discovery made by a man and is a twofold achievement'. Although this article addresses concerns that Hertha had voiced regularly (and in similar language) and makes mention of Marie Curie as a close friend, the piece was not in fact written by Hertha but by Dr. Marie Stopes (1880–1958), who was paid £3 for her contribution.[83] She was a supporter of women's suffrage and, like Hertha, had joined the short-lived Women's Freedom League, so it is probable that they were at least acquainted.

Hertha and the Hughes Medal

Despite the Royal Society's reluctance to admit women as fellows or add any trace of femininity to its public image, it did publish women's papers and award women the occasional grant. It awarded Hertha its Hughes Medal in 1906, the first woman to receive a medal in her own right (Marie Curie had been awarded the Davy Medal in 1903, along with her husband Pierre). The award of a prize or medal to a woman could be seen as an exceptional rarity and did not compromise the masculine character of the awarding body in the same way that admitting women as fellows would.[84] In fact, such awards could be justified within a discourse of gentlemanly patriarchy and endorsed, rather than challenged, men's control of science. John Perry and William Tilden (again) nominated Hertha for the award; Perry was now a member of the Royal Society council, the body in whose hands the election of medallists rested. He had nominated Hertha the previous year, but she had lost to Augusto Righi who was nominated by Larmor and Huggins. The year 1906 was only the third year that the Hughes Medal had been awarded; it was the last of a flurry of late-nineteenth-century bequests which included the Davy Medal (1869), Darwin Medal (1888), Buchanan Medal (1894) and Sylvester Medal (1897). After Professor Hughes's bequest in 1900, the Royal Society decided to accept no more as the process of finding suitable recipients was becoming too arduous.

Possibly still smarting from his defeat in 1902, in 1905 Perry argued Hertha's case on the grounds that her experimental work was 'so complex that many very clever scientific men abandoned the investigation'.[85] Perry highlighted Hertha's gender again when he and Tilden nominated her a second time in 1906, this time making a case three times longer than the earlier one and claiming that, due to the quality of her work, 'the exceptional step was taken of electing her a Member of that Institution (IEE) of which she still remains, after seven years, the only woman Member among a body of about 6000'.[86] Again Larmor nominated a candidate to stand against Hertha; Elihu Thomson was director of the Thomson-Houston Electrical Company in America where he had emigrated when he was five years old.[87] He was nominated for his work on the improvement of electric meters, the development of electric welding and metal working machinery, and alternating currents. He was not a strong candidate for the Hughes Medal (which was awarded for original discovery) and, in addition, he had previously been awarded the Rumford Medal. Even Larmor could only manage one paragraph arguing the case for Thomson and the vote went in Hertha's favour.

Conclusion

In *The Call*, Zangwill takes us inside the mind of the president of her fictional scientific society (a Huggins-type figure) as he ponders 'disgruntedly' a recent attempt to alter the charters to admit women '"... it would discredit the Society, reduce its meetings to frivolous social functions." The introduction of frivolity was the reason always advanced ... against the admission of women to masculine institutions'.[88] For Hertha sex had no bearing on the quality or kind of scientific work an individual pursued, as she made clear in an interview with the *Daily News* in 1919:

> I do not agree with sex being brought into science at all. The idea of 'woman and science' is entirely irrelevant. Either a woman is a good scientist, or she is not; in any case she should be given opportunities, and her work should be studied from the scientific, not the sex, point of view.[89]

At a time of profound social change when many voices, not least scientific ones, were seeking to define women's nature and role, this 'modern' notion of equality between the sexes was not a strong thread. Hertha was at odds with other scientific women who argued that women approached science differently and used this as a justification for feminine participation. Anthropologist Clémence Royer (like Hertha, a friend of Marie Curie) believed that women found their own, 'feminine' way to scientific truth and that their participation was essential to 'debias science'.[90] In 1913 H.J. Mozans made a plea, based on women's special way of reasoning, for male/female collaborations in science;[91] earlier mathematician Mary Somerville had expressed similar views, inspiring William Whewell's celebrated comment that 'there is sex in minds'. Within the suffrage movement too there was wide consensus that the vote was needed because women did things differently from men, and that women's complementary qualities were urgently required by the State. Hertha's relationship with the Royal Society was informed by a belief that women and men were no different when it came to scientific experiment and discovery; this was a conclusion that the whole culture of that learned body, which held fast to its masculine past, could not admit.

Despite this, the Royal Society had fewer qualms about accepting women's work for publication, offering them the occasional grant or medal, even a rare honorary fellowship; this the Society could achieve without jeopardising gendered spatial and ideological boundaries. Furthermore, acting as the judge of all scientific work, regardless of the sex of its producer, was important for the Society in maintaining its position as arbiter of excellence and gatekeeper of science, especially at a time when the growth of specialist societies was perceived as a threat to its status. Once it had been accepted, Hertha's paper on the electric arc was 'owned' by the Royal Society and she was required to seek the council's permission for publishing it in her 1903 book. In the interests of maintaining its position as the supreme guardian of science, the Royal Society was forced to recognise women's work; in the interests of assuring its high status in relation to other societies, it ensured that women, including Hertha Ayrton, were kept at the margins – never in the fellowship.

8
Conclusion

As I write, in the first decade of the twenty-first century, yet another debate is ongoing in the press about the merits of vocational versus academic education (particularly should diplomas replace A levels?) and groups such as WISE are still working to attract more women into science, engineering and mathematics.[1] What would Hertha Ayrton, who nearly achieved 'FRS' after her name in 1902, make of the fact that women are still a tiny minority within the Royal Society, in 2009 hovering around just 5% of the fellowship? There is little doubt that gender is still an issue to be confronted in any historical or social understandings of mathematics and science.

As much as anything, this book is a call to move gender from the periphery to a position closer (at least) to the theoretical centre of histories of mathematics and science. Crucially, this needs to extend beyond women and the perceived conflicts between femininity and the practise of any particular discipline. We need also to build on familiar understandings of the masculine nature of mathematics and science to delve deeper, and more precisely, into the nature of scientific or mathematical masculinity. How does it relate to femininity? How does it change over time and in different contexts? How does it affect the nature of practise and product? How does ideal or aspirational masculinity impact on the self-identity, and inclusion practices, of any particular discipline?

This small study has demonstrated that such questions are important to understanding the development of pure mathematics and practical science in the decades around 1900. At this time the role of women, along with Victorian ideas of gendered intellect, were adapting in response to the campaign for suffrage, eugenic thinking and the movement for women's higher education. In this fluid, developing context, differing concepts of femininity and masculinity were one of the pivotal issues around which the pure and the practical sought to establish their differing identities, legitimacy and moral integrity. Hertha, for example, sought to practise as a professional within a modern and meritocratic science that sought to accommodate commercialism and technology as well as research and discovery. Grace was representative of an older tradition which privileged notions of gifted intellect and an inherited aristocracy of talent and used these ideas to argue for the superiority of pure, abstract mathematics. Within each community, the meanings, signification and acceptance or rejection of women as participants varied: both women had to make negotiations which limited their opportunities and served to keep existing gender hierarchies in place.

In many ways, during the decades around 1900 femininity became representative of a reaction against the mechanisation and industrialisation of modern science and technology. Pure mathematics, too, positioned itself against this materialistic modernism by withdrawing from the world and privileging natural intellect. There were good reasons therefore for femininity and mathematics to exhibit an affinity at this time. Despite this, Grace's experience shows how gendered ideas of genius within mathematics served to exclude women – and cause them to exclude themselves – from its highest reaches. At the same time, practical science and engineering cultivated a virile, active identity centring on its provision of technology and 'manly' service to the world. Within this latter conception there was little room for a 'sedentary' femininity assumed to be more at home (literally and metaphorically) with bookwork and mathematics.

The antagonisms so apparent around 1900 were not new but reflected a long-standing chasm between Cartesian 'mind and body', or the more contemporary 'brain and hand'. These divisions, in England at least, were mapped on to notions of class. It is in keeping with notions of the higher moral worth of 'brain work' that lower middle-class Hertha, with her Jewish watchmaker father, should prefer practical science, and that quintessentially middle-class Grace, with her pretensions to an intellectual elite, should privilege the 'pure'. Although it is not suggested that these categories were rigid or did not overlap, these distinctions did cause tensions and inform debate within the scientific community and within bodies such as the Royal Society. The idea of a *learned* scientific method or 'attitude' (applied to life as well as to science) which was a central assumption of technical education, by necessity entailed a certain aesthetic loss. It is significant that 'abstraction' became the indicator of 'genius'. In Rayleigh's view, for example, 'facility for contrivance, backed by unflinching perseverance' were the characteristics of experimentalists, whereas the mathematical physicist was endowed with 'genius and insight'.[2] This hierarchy is based on the amount of abstract mathematics involved, a continuum in which pure mathematics was at the privileged extreme. In this dualism between the pure and the practical there are also echoes of the later 'two cultures' debate which is predicated on the division between science and the arts.[3] It is significant that within this distinction, where to place mathematics is always an issue.

The women's lives touched on here have not been aligned within any 'heroic' reclamation along the lines of the 'great men' view of history; instead, the focus has been on the more interesting question of how (and if) they were able to negotiate a place for themselves in their relative fields. Hertha encountered obstacles in being accepted as a working professional alongside her male peers. Grace was accepted more readily within the mathematical research community and, at the start of her career at least, facilitated a place in that community for her husband. Both Grace and Hertha's experiences illustrate the limitations of the 'mentor' model to explain fully the access of women to science in the late nineteenth century. Although undoubtedly useful in some instances – and certainly William Ayrton played this role at times for his wife – this model tends to be restrictive and present women as passive agents. Hertha was never the latter, arguing her case whenever necessary, pushing at doors and creating opportunities for herself. It must be remembered that all non-fellows of the Royal Society, male or female, needed a male mentor in the trivial sense that they required a male fellow to 'communicate' their papers.

There is little doubt that the decades surrounding 1900 were a high point for the participation of women in research mathematics, not only in Britain but in America and Germany too. This conclusion is supported by evidence (presented in Chapter 6) of the high proportion of women taking part II of the Cambridge mathematics tripos, the relatively high ratio of female fellows of the London Mathematical Society, and by women contributors to journals such as the *Educational Times*. That this participation reached a plateau and was not sustained after World War One is further evidence of the changes in configuration of mathematics that would benefit further research, as would cross-cultural comparisons.

As a last word, it is pleasing to note that in addition to their work and respective legacies to mathematics and science, Hertha and Grace are both remembered by academia. In 1925, Mrs. Charles Hancock (Ottilie Blind) endowed Girton College with a £3000 fellowship for the endowment of science in memory of her friend. Grace is remembered by the University of Wisconsin, USA, where a chair in mathematics has been established in her name.

Notes

Introduction

1 Marie Corelli, or Mary Mackay (1855–1924) was a successful romantic novelist. The quotation is from her pamphlet, *Woman or – Suffragette? A Question of National Choice* (London: C. Arthur Pearson, 1907).

2 Sally Ledger and Roger Luckhurst, eds, *The Fin de Siècle: A Reader in Cultural History c.1880–1900* (Oxford: Oxford University Press, 2000), p. xii.

3 A central text is Steven Shapin and Simon Schaffer, *Leviathan and the Air-pump: Hobbes, Boyle and the Experimental Life* (Princeton: Princeton University Press, 1985).

4 Otto Mayr, 'The science-technology relationship', in *Science in Context: Readings in the Sociology of Science*, ed. by Barry Barnes and David Edge (Milton Keynes: Open University Press, 1982), pp. 155–163.

5 *Nature*, November 8 1900, News, p. 28.

6 For example Alison Winter, 'A calculus of suffering: Ada Lovelace and the bodily constraints on women's knowledge in early Victorian England', in *Science Incarnate: Historical Embodiments of Natural Knowledge*, ed. by Christopher Lawrence and Steven Shapin (Chicago: University of Chicago Press, 1998), pp. 202–239 and Margaret Wertheim, *Pythagoras' Trousers: God, Physics and the Gender Wars* (London: Fourth Estate, 1997).

7 Carl B. Boyer, *A History of Mathematics* (New York: Wiley, 1968), pp. 649–650.

8 For the social construction of mathematics see David Bloor, 'Formal and informal thought', in *Science in Context: Readings in the Sociology of Science*, ed. by Barry Barnes and David Edge (Milton Keynes: Open University Press, 1982), pp. 117–124; *Math Worlds: Philosophical and Social Studies of Mathematics and Mathematics Education*, ed. by Sal Restivo, Jean Paul Bendegem and Roland Fisher (Albany: State University of New York Press, 1993).

9 Londa Schiebinger, *The Mind Has No Sex? Women in the Origins of Modern Science* (Cambridge, MA: Harvard University Press, 1989); Evelyn Fox Keller, *Reflections on Gender and Science* (New Haven: Yale University Press, 1985) and Ruth Watts, *Women in Science: A Social and Cultural History* (London: Routledge, 2007).

10 Marsha L. Richmond, '"A lab of one's own": The Balfour biological laboratory for women at Cambridge University, 1884–1914', in *History of Women in the Sciences: Readings from ISIS*, ed. by Sally Gregory Kohlstedt (Chicago: University of Chicago Press, 1999), pp. 235–268; Paula Gould, 'Women and the culture of University physics in late nineteenth-century Cambridge', *British Journal for the History of Science*, 30 (2) (1997), 127–149; Helena M. Pycior, Nancy G. Slack and Pnina G. Abir-am, eds, *Creative Couples in the Sciences* (New Brunswick, NJ: Rutgers University Press, 1996) and Pnina G. Abir-am and Dorinda Outram, eds, *Uneasy Careers and Intimate Lives: Women in Science, 1789–1979* (New Brunswick, NJ: Rutgers University Press, 1987).

11 Hertha always referred to herself as 'Mrs. Hertha Ayrton' instead of the more conventional 'Mrs. William Ayrton'.
12 For example Mary R.S. Creese, *Ladies in the Laboratory? American and British Women in Science, 1800–1900: A Survey of Their Contributions to Research* (London: Scarecrow Press, 1998); Bettye Anne Case and Anne M. Leggett, eds, *Complexities: Women in Mathematics* (Princeton: Princeton University Press, 2005); Marilyn Bailey Ogilvie, *Women in Science, Antiquity through the Nineteenth Century: A Biographical Dictionary with Annotated Bibliography* (Cambridge MA: MIT Press, 1991).
13 In particular Andrew Warwick, *Masters of Theory: Cambridge and the Rise of Mathematical Physics* (Chicago: Chicago University Press, 2003).

Chapter 1

1 Grace Chisholm Young commented of fellow mathematics student, Isabel Maddison, that she had '… the glamour of 'probably a wrangler' about her'. A 'wrangler' was the holder of a first-class pass in the Cambridge mathematical tripos. Liverpool University, Special Collections and Archives (LUSA), Papers of W.H. and G.C. Young (Young Papers) D140/12/22 (Grace's autobiographical notes).
2 Emily Davies, 'Special systems of education for women', in *The Education Papers: Women's Quest for Equality in Britain, 1850–1912*, ed. by Dale Spender (New York: Routledge and Kegan Paul, 1986), pp. 99–110 (p. 105).
3 Laurence Chisholm Young, *Mathematicians and Their Times* (Amsterdam: North-Holland, 1981), pp. 267–268.
4 Since the mid-eighteenth century Cambridge mathematics graduates were divided by class of degree into wranglers (first class), senior optimes (second class), junior optimes (third class) and pollmen (pass). For details of high offices held by senior wranglers see D.O. Forfar, 'What became of the Senior Wranglers?', *Mathematical Spectrum* (29) (1) (1996).
5 For teaching see Sheldon Rothblatt, *The Revolution of the Dons: Cambridge and Society in Victorian England* (Cambridge: Cambridge University Press, 1981); for relationship with the natural sciences tripos see David B. Wilson, 'Experimentalists among the mathematicians: Physics in the Cambridge Natural Sciences tripos, 1851–1900', *Historic Studies in the Physical Sciences*, 12 (2) (1982), 325–371; for bodily/mental training see Andrew Warwick, 'Exercising the student body: Mathematics and athleticism in Victorian Cambridge', in Lawrence and Shapin, pp. 288–326.
6 Fiona Erskine, '*The Origin of Species* and the science of female inferiority', in *Darwin's 'The Origin of Species': New Interdisciplinary Essays*, ed. by David Amigoni and Jeff Wallace (Manchester: Manchester University Press, 1995), pp. 95–121.
7 Sara A. Burstall, *Retrospect and Prospect: Sixty Years of Women's Education* (London: Longmans, Green & Co., 1933), p. 88.
8 Barbara Bodichon (1827–1891) was one of the founders of Girton College; she was involved in many feminist and educational projects including suffrage petitions, reform of the Married Women's Property Act, the Langham Place Group and the *English Woman's Journal*.

9   Numa Hartog, 1846–1871. Despite being senior wrangler, he was prevented from taking a fellowship by his inability to subscribe to the required religious test. He was a prominent figure in the movement for Jewish emancipation until his untimely death from smallpox. Joseph Jacobs and Goodman Lipkin, 'Hartog, Numa Edward', in *Jewish Encyclopedia.com* <http://www.jewish-encyclopedia.com>[accessed February 7 2005].

10 By 1925, Karl Pearson at University College London had illustrated with extensive statistics the inferiority of Jewish children with regard to intelligence, cleanliness of hair and tendency to breath through mouths, see David Albery and Joseph Schwartz, *Partial Progress: The Politics of Science and Technology* (London: Pluto, 1982), p. 176.

11 Judith Halberstam, 'Technologies of monstrosity: Bram Stoker's Dracula', in *Cultural Politics at the Fin de Siècle*, ed. by Sally Ledger and Scott McCracken (Cambridge: Cambridge University Press, 1995), pp. 248–266.

12 John A. Garrard, *The English and Immigration, 1880–1910* (Oxford: Oxford University Press, 1971), pp. 16–25.

13 M.C. Bradbrook, *'That Infidel Place': A Short History of Girton College, 1869–1969* (London: Chatto and Windus, 1969), p. 62.

14 Sharp, *Hertha Ayrton*, p. 54.

15 Ibid., p. 46.

16 Ibid., p. 136.

17 Burstall, *Retrospect and Prospect*, p. 78. Burstall reports that at Girton a 'legend' was current that Hertha was used by George Eliot (Marian Evans) as the model for 'Mirah' in *Daniel Deronda*; Hertha was known to Evans as the latter contributed to the funds Barbara Bodichon raised to support Hertha in her studies.

18 Linda Hunt Beckman, 'Leaving the tribal duck pond': Amy Levy, Jewish self-hatred and Jewish identity', *Victorian Literature and Culture*, 27 (1) (1999), 185–201 (p. 195). However, there are examples of male Jewish scientists who did not relinquish their religion yet achieved success in the same scientific environs as Hertha, for instance Raphael Meldola (1849–1915), Professor of Chemistry at Finsbury Technical College. Women, as Levy indicates, have special responsibility for family and domesticity within Judaism and this makes it more difficult for them to combine a Jewish identity with work beyond the family sphere.

19 Edith Ayrton Zangwill, *The Call* (London: Allen and Unwin, 1924). In this account Hertha/Ursula's family celebrate Christmas.

20 Sharp, *Hertha Ayrton*, pp. 27–28.

21 Ibid., p. 25.

22 This painting seems to be depicting the legend of Veronica, the pious woman of Jerusalem, who gave Jesus her handkerchief as he carried the cross.

23 LUSA, Young Papers, D140/12 (Grace's autobiographical notes).

24 Martha Vicinus, *Independent Women: Work and Community for Single Women, 1850–1920* (London: Virago, 1994), p. 39.

25 LUSA, Young Papers, D140/12/23 and 22 (Grace's autobiographical notes).

26 LUSA, Papers of Mrs. R.C.H. Tanner (Tanner Papers), D599/6 (Grace's autobiography). Grace records that she was unable to purchase tea or other items for personal use at Girton, 'the principle of the College being that all stu-

dents are on an equal footing and that richer students could not obtain privileges over poorer ones by paying for them'.

27 LUSA, Young Papers, D140/12/22-23 (Grace's autobiographical notes).

28 Sharp, *Hertha Ayrton*, p. 64.

29 Isabel Maddison (1869–1950). After becoming a wrangler in the mathematics tripos of 1892, Maddison joined Charlotte Angas Scott at Bryn Mawr to undertake doctoral study (supervised by Scott) where she won a fellowship to study abroad and joined Grace at Göttingen. Maddison returned to Bryn Mawr, completed her doctorate in 1896, and took a post on the staff where she remained until retirement.

30 Unlike Cambridge, Oxford did not formally recognise the women's halls and imposed no residency requirement on women taking their examinations.

31 This consisted of a watch spring fastened over the artery on the wrist with a marker (paint brush) attached to the other end of the spring which oscillated with the pulse. When a paper was pulled across the marker/paint brush at a uniform rate the pulse was recorded.

32 Burstall, *Retrospect and Prospect*, pp. 87–88. Emily Davies wrote in commiseration 'I have no doubt that in spite of this misfortune, you have a useful and honourable career before you ...'.

33 Ernest Hobson (1856–1933) was a research mathematician as well as a well-regarded coach; he was in the forefront, with Grace Chisholm Young and William Henry Young, in introducing the theory of functions to Cambridge.

34 Philippa Fawcett (1868–1948) beat the senior wrangler by 400 marks or 13% and went on to be placed in the first class of the tripos Part II the following year; she was then awarded a research scholarship at Newnham after which she became a college lecturer there. She published only one mathematical research paper, leaving Newnham in 1902 to pursue a career in educational administration. See Stephen Siklos, *Philippa Fawcett and the Mathematical Tripos* (Cambridge: Newnham College, 1990).

35 Ibid., pp. 30–31.

36 Rita McWilliams-Tullberg, *Women at Cambridge: A Men's University – Though of a Mixed Type* (London: Gollancz, 1975), p. 124.

37 Jean Barbara Garriock, *Late Victorian and Edwardian Images of Women and Their Education in the Popular Periodical Press with Particular Reference to the Work of L.T. Meade* (unpublished doctoral thesis, University of Liverpool, 1997), p. 40.

38 S.A. Burstall, 'The Place of Mathematics in Girls' Education', *The Mathematical Gazette*, 6 (96) (1912), pp. 203–213, p. 205. Burstall was then Headmistress of Manchester High School for Girls.

39 Quoted in Carol Dyhouse, 'Good wives and little mothers: Social anxieties and the schoolgirl's curriculum, 1890–1920', *Oxford Review of Education*, 3 (1) (1977), 21–35 (p. 25).

40 Geoffrey Howson, *A History of Mathematical Education in England* (Cambridge: Cambridge University Press, 1982), pp. 173–174.

41 Patricia Vertinsky, 'Exercise, physical capability, and the eternally-wounded woman in late nineteenth century North America', *Journal of Sport History*, 14 (1) (1987), 7–27.

42 Barbara Stephen, *Emily Davies and Girton College* (London: Constable, 1927), p. 290.

43 Mrs. Henry Sidgwick, *Health Statistics of Women Students of Cambridge and Oxford and of Their Sisters* (Cambridge: Cambridge University Press, 1890), p. 66. The collection of health statistics on students was a common response by women's colleges to medical anxieties; many studies of this kind, often generating reassuring results, were carried out in the United States, see Vertinsky, p. 20.

44 Margaret Burney Vickery, *Buildings for Bluestockings: The Architectural and Social History of Women's Colleges in Late-Victorian England* (Newark: University of Delaware Press, 1999), pp. 152–155.

45 LUSA, Young Papers, D140/6/38.

46 Edwin A. Abbot, *Flatland: A Romance of many Dimensions*, 6th edn (Oxford: Blackwell, 1950).

47 Sheldon Rothblatt, *The Revolution of the Dons: Cambridge and Society in Victorian England* (Cambridge: Cambridge University Press, 1981), p. 234.

48 This was a problem for mathematical women even in the 1920s. Mary Cartwright, who gained a first at Oxford and went on to contribute research on chaos theory and become the first female president of the London Mathematical Society, recalled that she came up to university knowing she was ill-prepared. See James Tattersall, Shawnee McMurran and Mary L. Cartwright, 'An Interview with Dame Mary L. Cartwright, D.B.F., F.R.S.', *The College Mathematics Journal*, 32 (4) (2001), 242–254 (p. 247).

49 Burstall, *Retrospect and Prospect*, pp. 83–84.

50 Warwick, *Masters of Theory*, p. 281.

51 Sarah Marks, 'Abstracts from letters to Barbara Bodichon' (January 30 1880), *Girton Review*, Michaelmas Term (1927), 8–11 (p. 10).

52 LUSA, Young Papers, D140/12/22. Bennett did indeed become senior wrangler that year and was indeed beaten by Fawcett who was ineligible to take the title herself due to her sex. Webb did not carry out his threat to emigrate and eventually retired from coaching at Cambridge in 1902.

53 Rev. Dr. R.S. Franks, 'Mr. Robert Webb', *The Times*, August 5 1936, Obituaries, p. 14.

54 Philip G. Hamerton, *The Intellectual Life* (London: Macmillan, 1911), pp. 261–264.

55 Burstall, *Retrospect and Prospect*, p. 88.

56 Arthur Cayley (1821–1895) Sadleirian Professor of Mathematics at Cambridge, President of the London Mathematical Society 1868–70.

57 Patricia C. Kenschaft, 'Charlotte Angas Scott, 1858–1931', *College Mathematics Journal*, 18 (2) (1987), 98–110 (p. 102).

58 Bradbrook, p. 32.

59 LUSA, *Young Papers*, D140/12/22–23.

60 Sarah Marks, 'Abstracts from letters to Barbara Bodichon', p. 10.

61 LUSA, Young Papers, D140/34/55 (Grace's autobiographical notes).

62 Ibid., D140/6/160.

63 Edward J. Routh was renowned as a 'wrangler master'. He coached more than 600 students between 1855 and 1888 including 27 senior wranglers. A.T. Fuller, 'Routh Edward John (1831–1907)', *Oxford Dictionary of National Biography* (Oxford: Oxford University Press, 2004), <http://www.oxforddnb.com/view/article/35850> [accessed February 9 2005].

64 A.R. Forsyth, 'Old Tripos Days at Cambridge', *Mathematical Gazette*, 29 (1935), 162–179 (p. 173).

65 Franks, 'Mr. Robert Webb'.
66 Sharp, *Hertha Ayrton*, p. 56.
67 LUSA, Young Papers, D140/6/1-32 (Grace's autobiographical notes).
68 Margaret E. Tabor, 'Philippa Garrett Fawcett, 1887–1902', *Newnham College Roll Letter* (January 1949), 46–51 (p. 47).
69 Forsyth, 'Tripos Days', p. 174.
70 LUSA, Young Papers, D140/12/5.1 (Grace's autobiographical notes).
71 Ibid., D140/12/23.
72 LUSA, Tanner Papers, D599/6 (Grace's autobiography).
73 Tattersall et al., 'An Interview', p. 249.
74 Vickery, pp. 12–39 (p. 21).
75 Garriock, p. 85 (referring to women at Newnham College in 1895).
76 Charlotte Angas Scott, 'Paper Read before the Mathematical Club at Girton College, May Term, 1893', *Girton Review*, 36 (1894), 1–4 (p. 2).
77 Alice Gardner, *A Short History of Newnham College, Cambridge* (Cambridge: Bowes, 1921), p. 77.
78 Sarah Marks, 'Abstracts from letters to Barbara Bodichon', p. 9.
79 Tullberg, pp. 102–103.
80 'E.W. Hobson', Obituary Notices of Royal Society of London, (3) (1934), 239.
81 Young, *Mathematicians and their Times*, p. 278. It should be noted that Philippa's mother, Millicent Fawcett, was not a militant feminist but a supporter of the law-abiding, conciliatory wing of the suffrage movement.
82 Quoted in Patricia C. Kenschaft, 'Charlotte Angas Scott, 1858–1931', in *Women of Mathematics*, ed. by Louise S. Grinstein and Paul J. Campbell (Westport, Conn: Greenwood, 1987), pp. 193–203 (p. 197).
83 Warwick, *Masters of Theory*, pp. 176–226.
84 Paul Atkinson, 'Fitness, Feminism and Schooling', in *The Nineteenth Century Woman: Her Cultural and Physical World*, ed. by Sara Delamont and Lorna Duffin (London: Croom Helm, 1978), pp. 92–133 (pp. 106–107).
85 Jennifer A. Hargraves, '"Playing like Gentlemen while behaving like Ladies": Contradictory features of the formative years of women's sport', *British Journal of Sports History*, 2 (1985), 40–52 (p. 43).
86 Bradbrook, p. 104.
87 Warwick, 'Exercising the Student Body', p. 299.
88 Lord Rayleigh, Robert John Strutt, *Life of Sir J.J. Thomson* (Cambridge: Cambridge University Press, 1942), p. 10.
89 G.H. Hardy, *A Mathematician's Apology, with a forward by C.P. Snow*, 2nd edn (Cambridge: Cambridge University Press, 2001), p. 18.
90 University College London, Library Manuscripts Room, papers and correspondence of Sir Francis Galton, 1822–1911, 196/9 (letter Grace Chisholm Young to Francis Galton, May 29 1909).
91 For example see Hardy, pp. 22–24.
92 This is a reference to Cambridge algebraist and Professor of Mathematics Arthur Cayley – 'a wonder in pure mathematics': Forsyth, 'Old Tripos Days', pp. 162–163.
93 Sir Stafford Northcote, 1873, quoted in Burstyn, p. 73.
94 Young, *Mathematicians and their Times*, p. 267. See also Garriock, p. 84.
95 Testimonial, Donald Macalister, *Archives of Manchester High School for Girls*, L 1886 1.

96 I. Grattan-Guinness, 'University mathematics at the turn-of-the-century: Unpublished recollections of W.H. Young', *Annals of Science*, 28 (4) (1972), 367–384 (p. 373).

97 LUSA, Young Papers, D140/6/55 and D140/2/2.1.

98 William J. Ashworth, 'Memory, efficiency and symbolic analysis: Charles Babbage, John Herschel and the Industrial Mind', *ISIS*, 87 (4) (1996), 629–653.

99 William Whewell (1794–1866) of Trinity College, Cambridge, a leading figure in early-mid nineteenth century science, mathematics and philosophy.

100 Janet Howarth, '"In Oxford but ... not of Oxford": The Women's Colleges', in *The History of the University of Oxford* (Vol. VII, Nineteenth-Century Oxford, Part 2) ed. by M.G. Brocks and M.C. Curthoys (Oxford: Clarendon Press, 2000), pp. 237–307 (pp. 239–240).

101 Ibid., pp. 257–258.

102 Ibid., pp. 282–283.

103 Pauline Adams, *Somerville for Women: An Oxford College 1897–1993* (Oxford: Oxford University Press, 1996), p. 34.

104 Ibid., p. 53.

105 Suggested figures for Oxford from Howarth, p. 282, who notes that these figures are indicative only as they do not include home students and only around half of the students at the women's halls sat examinations.

106 Carol Dyhouse, *No Distinction of Sex? Women in British Universities, 1870–1939* (London: UCL Press, 1995), pp. 12–13.

107 Oxbridge women 'graduates' in all but name also travelled to Trinity College, Dublin, to gain formal degrees. In 1904 the Board of Trinity College passed a grace to allow women from Cambridge and Oxford colleges to apply for degrees in the three-year period 1904–1907, after which time Trinity would have its own women graduates. Over 700 women came to Dublin for this purpose, often taking out BA and MA degrees on the same day. They travelled by sea and came to be known as 'Steamboat Ladies'. See Deirdre Raftery and Susan M. Parkes, eds, *Female Education in Ireland 1700–1900: Minerva or Madonna* (Dublin: Irish Academic Press, 2007), p. 131. To 1914, only four women are identified as mathematics graduates of Trinity College, the first earning their BA degrees in 1912 (Davis Archive).

108 Negley Harte, *The University of London 1836–1986: An Illustrated History* (London: The Athlone Press, 1986), p. 128.

109 Charlotte Angas Scott was supervised by Arthur Cayley at Cambridge. The first woman to receive a DSc was Sophia Bryant, in Moral Sciences, in 1884.

110 *London University Calendar*, 1880–1928 (London: University of London, 1878–1914+); Davis Archive.

111 Ada Lovelace (1815–1852) was trained in calculus by Augustus De Morgan (1806–1871); she formed a long friendship with Charles Babbage and translated/expanded a thesis on his analytical engine (a conceptual forerunner of the computer). See Benjamin Woolley, *The Bride of Science: Romance, Reason and Byron's Daughter* (London: Macmillan, 1999).

112 Geoffrey Howson, *A History of Mathematical Education in England* (Cambridge: Cambridge University Press, 1982), p. 177.

113 *London University Calendar*.

114 *Royal Holloway College Calendar*, 1897–1914.
115 Beverly Lyon Clark, 'Of Snarks and Games … and Publishing', *Children's Literature Association Quarterly*, 16 (2) (1991), 91–92 (p. 91).

Chapter 2

1 LUSA, Young Papers, D140/8/60 (Grace to mother, April (n.d.) 1895).
2 Grace Chisholm, 'On the curve and its connection with an astronomical problem', *Royal Astronomical Society Monthly Notices*, 57 (1895–7), 379–387.
3 LUSA, Young Papers, D140/14/6 (This is a letter/personal account written by Grace for her daughter Cecily on Christmas day 1920. Cecily also studied mathematics at Girton College, Cambridge).
4 LUSA, Young Papers, D140/6/392 (Grace to William Henry Young, November (n.d.) 1900).
5 Bennett's essay was on 'The residues of powers of numbers for any composite real modulus'; see June Barrow-Green, 'A Corrective to the Spirit of too Exclusively Pure Mathematics': Robert Smith (1689–1768) and his Prizes at Cambridge University', *Annals of Science*, 56 (3) (1999), 271–316, p. 309.
6 Siklos, p. 39. Bennett lectured on geometry and had a keen interest in the geometry of mechanisms.
7 Charlotte Angas Scott (1858–1931) studied mathematics at Girton but, because Cambridge did not award degrees to women, her BSc and DSc were awarded by London University. She taught at Girton 1880–84 before travelling to Bryn Mawr where she became Professor of Mathematics until 1917 and President of the American Mathematical Society in 1905. Isabel Maddison (1869–1950) was Angas Scott's first doctoral student and later joined the staff at Bryn Mawr. Patricia C. Kenschaft, 'Charlotte Angas Scott, 1858–1931', *College Mathematics Journal*, 18 (2) (1987), 98–110.
8 Mary (May) Winston (1869–1959) became the first American woman to receive a PhD in mathematics from Germany when she completed her studies soon after Grace.
9 Christine Ladd Franklin (1847–1930) worked on algebraic logic and, later, published research on colour vision; she eventually held teaching posts at Johns Hopkins and Columbia Universities but devoted much of her life to furthering postgraduate opportunities for women in mathematics. In 1926, Johns Hopkins eventually awarded her the PhD she had been denied forty-four years earlier; she was then aged 78. Judy Green and Jeanne LaDuke, 'Women in the American Mathematical Community: The Pre-1940 Ph.D.'s', *The Mathematical Intelligencer*, 9 (1) (1987), 11–23 (p. 13).
10 Winifred Edgerton Merrill (1862–1951) taught mathematics in various schools. Hers was the first degree given to a woman by Columbia University. For tables detailing the names of early women mathematics PhDs and awarding institutions, see Green and LaDuke, pp. 15–19.
11 Barbara Miller Solomon, *In the Company of Educated Women: A History of Women and Higher Education in America* (New Haven: Yale University Press, 1985), pp. 80–81.
12 George Weisz, *The Emergence of Modern Universities in France, 1863–1914* (Princeton NJ: Princeton University Press, 1983), p. 235.

13   Ibid., p. 246. The figures for foreign women students in 1914 are 469 (Medicine); 138 (Science) and 1,033 (Letters).
14   Susan Quinn, *Marie Curie: A Life* (Cambridge MA: Da Capo Press, 1995), p. 96.
15   Green and LaDuke, p. 11; Parshall and Rowe, pp. 239–253.
16   Renate Tobies, 'In Spite of Male Culture: Women in Mathematics', in Rachel Camina and Lisbeth Fajstrup, eds, *European Women in Mathematics: Proceedings of the 9th General Meeting* (New York: Hindaw, 2007), pp. 25–35, p. 32.
17   Emmy Noëther (1882–1935) contributed mathematical foundations to Einstein's theory of relativity and made fundamental contributions to modern algebra. Hilbert and Klein tried to have her appointed to the staff at Göttingen between the wars but were unsuccessful so Noëther taught classes which were 'nominally' Hilbert's lectures, without pay or position. She was dismissed by the Nazis in 1933 because she was Jewish and fled to the USA as a visiting professor at Bryn Mawr. See Lynn M. Osen, *Women in Mathematics* (Cambridge, MA: MIT Press, 1974), p. 151.
18   Althoff (1839–1908) managed the Prussian universities from 1882–1907, leading them through a period of expansion. It was through his vision and support of Felix Klein that Göttingen became such an important centre of mathematics. David E. Rowe, '"Jewish Mathematics" at Göttingen in the era of Felix Klein', *ISIS*, 77 (3) (1986), 422–449, p. 427 and p. 435.
19   An extra hurdle for women in Germany was the requirement for another qualification in addition to a doctorate for professorial positions. The 'Habilitation' was in effect a second doctorate obtained via becoming a 'senior assistant' and taking an examination open only to men, as was made clear in a law specifically excluding women passed in 1907 (repealed 1920). Tobies, p. 28.
20   Sandra L. Singer, *Adventures Abroad: North American Women at German-Speaking Universities, 1868–1915* (Connecticut: Praeger, 2003). This was to counter an influx of Russian women to Germany after Russia recalled its women from the University of Zurich.
21   See Ann Hibner Koblitz, *A Convergence of Lives: Sofia Kovalevskaia: Scientist, Writer, Revolutionary* (New Brunswick N.J.: Rutgers University Press, 1993), pp. 121–123. Koblitz recounts how one of the papers produced by Kovalevskaia was highly significant in the field of partial differential equations. Kovalevskaia died of pneumonia in 1891, aged forty-one.
22   Sara Delamont has identified two available lifestyles for graduates of the new colleges for women, the 'celibate' career woman who entered teaching and 'the learned wife': Delamont, p. 142.
23   Angas Scott supervised six women doctoral candidates and thanks to her leadership 'women were far more active in the American mathematical community than they were later, earning 14% of the doctorates in mathematics awarded before 1940 as compared to only 5% in the 1950s': Kenschaft, p. 105.
24   LUSA, Young Papers, D140/2/2.1.
25   Ibid., D140/6/329 (This is a letter from Grace to Frances de Grasse Evans, a close friend from Girton).
26   LUSA, Young Papers, D140/6/34-46 (Grace's Göttingen correspondence, c. November 1893).

27 LUSA, Young Papers, D140/6/267-328 (William Henry Young to Grace's mother, March 19 1899. The occasion referred to is a dinner held in honour of Felix Klein by the professors of the Mathematical and Physical Sciences at the University of Turin).

28 See Ute Frevert, *Women in German History: From Bourgeois Emancipation to Sexual Liberation*, trans. by Stuart McKinnon-Evans, Terry Bond and Barbara Norden (Oxford: Berg, 1986), pp. 107–137.

29 LUSA, Young Papers, D140/8/1-321 (courtship correspondence, October (n.d.) 1893).

30 Constance Reid, *Hilbert* (London: Allen and Unwin, 1970), pp. 139–140.

31 LUSA, Young Papers, D140/4/2 (Grace to Young, January 10 1901).

32 Ibid., D140/6/43a (Grace to friend, October (n.d.) 1893). The young lecturer was Ernst Ritter, Klein's assistant.

33 Sophia Smith Collection, Smith College, Massachusetts, USA. *Mary Frances Newson Winston Papers*, (hereafter 'Winston Papers'), correspondence October 15 1893.

34 David E. Rowe, 'Making mathematics in an oral culture: Göttingen in the era of Klein and Hilbert', *Science in Context*, 17 (1/2) (2004), 85–129 (p. 96).

35 LUSA, Young Papers, D140/6/34-46 (Göttingen correspondence, October (n.d.) 1893).

36 Winston Papers, personal recollections, pp. 6–7.

37 James C. Albisetti, *Schooling German Girls and Women: Secondary and Higher Education in the Nineteenth Century* (Princeton: Princeton University Press, 1988), pp. 131–132.

38 Katharina Rowold, 'The many lives and deaths of Sofia Kovalevskaia: Approaches to women's role in scholarship and culture in Germany at the turn of the twentieth century', *Women's History Review*, 10 (4) (2001), 603–628.

39 Karen Hunger Parshall and David E. Rowe, *The Emergence of the American Mathematical Research Community, 1876–1900: J.J. Sylvester, Felix Klein and E.H. Moore* (Providence RI: American Mathematical Society, 1991), p. 244.

40 Quoted Tobies, p. 30. Despite this, Tobies has found that at the beginning of the twentieth century in Germany mathematics was 'one of the most popular courses of studies for women' (p. 25).

41 Parshall and Rowe, pp. 123–124. However Planck could make exceptions: he eventually permitted physicist Lise Meitner to attend his lectures at Berlin in the early 1900s 'on a trial basis and always revocably ... (but) ... I must hold fast to the idea that such a case must always be considered as exception, and in particular that it would be a great mistake to establish special institutions to induce women into academic study, at least not into pure scientific research. Amazons are abnormal, even in intellectual fields.' See Ruth Lewin Sime, *Lise Meitner: A Life in Physics* (California: University of California Press, 1996), pp. 25–26.

42 UCL, Galton Papers, 344/2 (Grace to Galton, May 29 1909).

43 Albisetti, pp. 234–235.

44 LUSA, Young Papers, D140/8/1–321 (Göttingen correspondence, November (n.d.) 1893).

45 Parshall and Rowe, p. 240.

46 Parshall and Rowe, p. 244.

47 LUSA, Young Papers, D140/12/22. This remained unpublished.

48 UCL, Galton Papers, 344/2 (Grace to Galton, May 9 1909).

49 Greta Jones, *Social Hygiene in Twentieth-century Britain* (London: Croom Helm, 1986), especially pp. 18–19.

50 '... the eugenic theory of society, as elaborated by Galton, is a way of reading the structure of social classes on to nature': Donald A. Mackenzie, *Statistics in Britain 1865–1930: The Social Construction of Scientific Knowledge* (Edinburgh: Edinburgh University Press, 1981), p. 18.

51 Paul Weindling, *Health, Race and German Politics between National Unification and Nazism, 1870–1945* (Cambridge: Cambridge University Press, 1989), p. 9.

52 Quoted in Ellen Kennedy, 'Nietzsche: Women as Untermensch', in *Women in Western Political Philosophy: Kant to Nietzsche*, ed. by Ellen Kennedy and Susan Mendus (Brighton: Wheatsheaf, 1987), pp. 179–201 (p. 185).

53 Despite this misogyny, in Germany and England some feminists subverted Nietzsche's stricture to follow instinct and found an egalitarianism in his call to 'be what you are': Hinton R. Thomas *Nietzsche in German Politics and Society: 1890–1918* (Manchester: Manchester University Press, 1983), pp. 80–88; Lucy Delap, 'The Superwoman: Theories of gender and genius in Edwardian Britain', *Historical Journal*, 47 (1) (2004), 101–126.

54 LUSA, Tanner Papers, D599/16 (Grace's notes). As this poem is handwritten in Grace's personal notebooks, it was presumably either written or endorsed by her.

55 LUSA, Young Papers, D140/12/67.

56 Ibid., D140/6/318a.

57 For example, Reid, *Hilbert;* and Herbert Meschkowski, *Ways of Thought of Great Mathematicians: An Approach to the History of Mathematics*, trans. by John Dyer-Bennet (San Francisco: Holden Day, 1964).

58 Reid, p. 46.

59 LUSA, Young Papers, D140/8/1–321 (Göttingen correspondence, October (n.d.) 1893).

60 Reid, p. 89.

61 Winston Papers, correspondence, October 17 1893.

62 Ibid., p. 102.

63 LUSA, Tanner papers, D599/16.

64 Social prescriptions did not allow women to duel as duelling was intimately connected to masculinity. Duelling was also popular in France from around 1860 to World War One where it was used as a strategy to retain certain cultural and political arenas as wholly male. See Robert A. Nye, *Masculinity and Male Codes of Honor in Modern France* (Oxford: Oxford University Press, 1999), pp. 172–215.

65 Patricia M. Mazón, *Gender and the Modern Research University: The Admission of Women to German Higher Education, 1865–1914* (California: Stanford University Press, 2003), p. 36.

66 In a thought-provoking article, David E. Rowe has highlighted the social aspects of the production and teaching of mathematics at Göttingen; despite this, mathematicians gained great, individual reputations and this process may have been encouraged by the collective, oral ('shared') nature of mathematical research which created an environment within which individual talent and creativity could be displayed – and admired. Rowe, 'Making mathematics in an oral culture'.

67 Parshall and Rowe, p. 190.
68 Rowe, 'Making mathematics in an oral culture', p. 97.
69 LUSA, Young Papers, D140/12/22 (Grace's autobiographical notes).
70 For example W.H. Young, 'Reply', *Messenger of Mathematics*, 42 (1913), 113.
71 Reid, p. 92.
72 Mary L. Cartwright, 'Grace Chisholm Young', *Girton Review* (Spring Term, 1944), 17–19 (p. 19).
73 W.H. Young and Grace Chisholm Young, *The Theory of Sets of Points* (Cambridge: Cambridge University Press, 1906), Preface.
74 LUSA, Young Papers, D140/7/3 (Grace's notes).
75 Ibid., D140/14/1.
76 Meschkowski, pp. 94–95.
77 Ethel Sidgwick, *Mrs Henry Sidgwick: A Memoir by Her Niece* (London: Sidgwick and Jackson, 1938), p. 65.
78 Hardy, *Apology*, pp. 119–121. Hardy (1877–1947) was an analyst and friend of the Youngs; he was a professor of mathematics at both Cambridge and Oxford Universities.
79 Ibid., pp. 84–85. This hierarchy of pure over applied was given even more significance by those (unlike Hardy) with devout religious belief; for Hilda Hudson, Cambridge wrangler and respected researcher, pure mathematics was 'a branch of theology' that offered 'direct contact with God'. Hudson was writing in 1925, after the trauma of World War One; see Hilda P. Hudson, 'Mathematics and Eternity', *The Mathematical Gazette*, XII (174) 1925, 265–270 (pp. 265–266).
80 Cynthia Cockburn, 'Technology, production and power', in *Inventing Women: Science, Technology and Gender*, ed. by Gill Kirkup and Laurie Smith Keller (Milton Keynes: Open University Press, 1992), pp. 196–211 (p. 199).
81 Lewis Pyenson, *Neohumanism and the Persistence of Pure Mathematics in Wilhelmian Germany* (Philadelphia: American Philosophical Society, 1983), p. 54. Unsurprisingly, the engineers were unimpressed with Klein's comments and objected that theoreticians did not understand technical education.
82 Renate Tobies, 'Why a Felix Klein prize?', *ECMI Newsletter*, 27 (March 2000) http://www.mafy.lut.fi/EcmiNL/ [accessed March 24 2009]. See also Rowe, 'Klein, Hilbert and the Göttingen mathematical tradition', pp. 202–204.
83 W.H. and Grace Chisholm Young, preface.
84 LUSA, Young Papers, D140/12/1-12 (Grace's notes). The term 'Queen of the Sciences' was a common characterisation which had been used since before the seventeenth century. See Schiebinger, pp. 119–159 (p. 146).
85 Young, *Mathematicians and their Times*, p. 246.
86 Hardy, *Apology*, p. 135.
87 Susan P. Casteras, 'The cult of the male genius in Victorian painting,' in *Rewriting the Victorians: Theory, History and the Politics of Gender*, ed. by Linda M. Shires (New York: Routledge, 1992), pp. 116–146.
88 Hardy described himself as such in a 1942 letter to Grace Chisholm Young: LUSA, Young Papers, D140/9/64.
89 Christine Battersby, *Gender and Genius: Towards a Feminist Aesthetics* (London: The Woman's Press, 1994), pp. 4–5.

90 Delap, 'The Superwoman', p. 103.
91 Ed. Cohen, *Talk on the Wilde Side: Toward a Genealogy of Discourse on Male Sexualities* (New York: Routledge, 1993), p. 32.
92 LUSA, Young Papers, D140/14/6 (Grace's notes for her daughter, 1920).
93 Ashworth, 'Memory, efficiency and symbolic analysis', pp. 646–649.
94 Bloor, pp. 117–124.
95 Bertrand Russell, 'The study of mathematics', in *Mysticism and Logic, and Other Essays* (London: Penguin, 1953), pp. 60–61 (p. 60).
96 Battersby, p. 148.
97 Martina Kessel, '"The 'Whole Man'": The longing for a masculine world in nineteenth-century Germany', *Gender and History*, 15 (1) (2003), 1–31 (p. 2).
98 Reid, p. 46.
99 Wertheim, p. xv.
100 For a discussion of the problematic nature of proof at this time see Jeremy J. Gray, 'Anxiety and abstraction in nineteenth-century mathematics', *Science in Context*, 17 (1/2) (2004), 23–47 (pp. 27–29).

Chapter 3

1 The ETU's use of an abstract female figure to represent electricity and their organisation adhered to longstanding traditions; see Marina Warner, *Monuments and Maidens: The Allegory of the Female Form* (London: Weidenfeld and Nicolson, 1985).
2 Julie Wosk, *Women and the Machine: Representations from the Spinning Wheel to the Electronic Age* (Baltimore: Johns Hopkins Press, 2001), p. 17.
3 Ibid., 'The electric Eve', pp. 68–88. Wosk has traced such images of technology and women back to the mid-eighteenth century.
4 Ibid., pp. 18–19.
5 Graeme Gooday, 'Faraday reinvented: Moral imagery and institutional icons in Victorian electrical engineering', *History of Technology*, 15 (1993), 190–205.
6 Sharp, *Hertha Ayrton*, p. 182 (Hertha's comments in an interview with the *Daily News*, July 16 1919).
7 Ibid., p. 144. Hertha's experiments could be spectacular in the context of a lecture, as she explained in *Nature*, it was impossible for her to use an ordinary enclosed arc lamp as these used currents of up to only 8 amperes, 'whereas to test my theory it was necessary to employ currents up to 40 amperes. Accordingly, I constructed little electrical furnaces of different kinds …', Hertha Ayrton, 'The reason for the hissing of the electric arc', *Nature*, 60 (July 17, 1899), 302–305 (p. 303).
8 Institution of Electrical Engineers (IEE), Library and Archives, memoirs of A.P. Trotter, pp. 569–591. Trotter had been a president of the IEE.
9 London Guildhall Library (LGL), Department of Manuscripts, Records of Finsbury College, 29,973 Hertha is listed as 'Sarah Marks'.
10 Three patents were taken out during 1883 and 1884, the first in conjunction with her cousin Ansel Leo.
11 Tattersall and McMurran, p. 94 (advertising for 'Marks' Patent Line Divider).
12 Sharp, *Hertha Ayrton*, pp. 108–109 (from *Academy Magazine*).

13 A.D. Morrison-Low, 'Women in the nineteenth-century scientific instrument trade', in *Gender and Scientific Enquiry, 1780–1945*, ed. by M. Benjamin (Oxford: Basil Blackwell, 1991), pp. 89–117.

14 LGL, 21,986 (Prof. J. Perry, 'Address to Finsbury Technical College and Old Students' Association, 1888'). John Perry (1850–1920) taught Mechanics and Applied Mathematics at Finsbury, transferring to the Central Technical College in 1884. With William Edward Ayrton (Hertha's husband) he invented many electrical measuring instruments and worked on railway electrification and in other areas. Perry was elected to the Royal Society in 1885.

15 W.H. Brock, 'Building England's first technical college: The laboratories of Finsbury Technical College, 1878–1926', in *The Development of the Laboratory: Essays on the Place of Experiment in Industrial Civilization*, ed. by F.A.J.L. James (Basingstoke: Macmillan, 1989), pp. 154–170 (p. 166).

16 Graeme Gooday has demonstrated the processes by which this new academic space became accepted by industry as a viable engineering training venue: Graeme J.N. Gooday, 'Teaching telegraphy and electrotechnics in the physics laboratory, William Ayrton and the creation of an academic space for electrical engineering in Britain, 1873–1884', *History of Technology Journal*, 13 (1991), 73–111.

17 Cambridge, Girton College Library and Archive (GCLA), unpublished MS by Joan Mason, 'Matilda Chaplin Ayrton (1846–1883), William Edward Ayrton (1847–1908) and Hertha Ayrton (1854–1923)', (1994), p. 6.

18 LGL, 21,980 (Finsbury Park evening examination results).

19 W.E. Ayrton, *Practical Electricity: A Laboratory and Lecture Course* (London: Cassell, 1900), p. xii.

20 Gooday, 'Teaching telegraphy and electrotechnics', pp. 102–103.

21 For example, the collection of essays in Pycior and others, eds, discussed in the introduction.

22 Sharp, *Hertha Ayrton*, p. 114 (letter from Hertha to her mother).

23 UCL, Galton Papers, 196/9 (correspondence W.E. Ayrton to Galton, February 17 1907).

24 Matilda Chaplin Ayrton was one of the 'Edinburgh Seven' with Sophia Jex-Blake. She travelled to Paris to study for a medical degree, gained a Certificate in Midwifery from the London Obstetric Society and, while Ayrton was working in Tokyo, introduced Western midwifery practices into Japan: GCLA, Mason MS.

25 LGL, 21,868/10 ('Funeral Notice, W.E. Ayrton', *Morning Post*, November 13th, 1908).

26 Israel Zangwill, 'Professor Ayrton', *The Times*, November 11, 1908, Letters to Editor, p. 15. Zangwill, the Jewish writer, was a son-in-law of William and Hertha Ayrton and a member of the Men's League for Women's Suffrage.

27 Thomas Mather (1856–1937) had been William Ayrton's Chief Assistant at Finsbury and the Central, where he succeeded Ayrton as Professor. On Hertha's death he rushed to counter accusations, made by professor of chemistry Henry E. Armstrong, that Hertha had been unoriginal and dependent on her husband: Thomas Mather, 'Mrs Hertha Ayrton', *Nature*, 112 (December 29, 1923), 939.

28 Maurice Soloman, 'Review of Evelyn Sharp's *Hertha Ayrton, 1854–1923: A Memoir*', *Central Gazette*, 23 (59) (1926), 70–72 (p. 72).

29 Michael Argles, *South Kensington to Robbins: An Account of English Technical and Scientific Education since 1851* (London: Longman, 1964), p. 28.

30 Hilda Phoebe Hudson (Newnham College, Cambridge) who was placed equal to the seventh wrangler in 1904, had a twin brother with equal mathematical talent who was killed in a climbing accident just as he was embarking on mathematical research: M.D.K., 'Hilda Phoebe Hudson, 1881–1965', *Newnham College Roll Letter* (1966), 53–54.

31 '… elite masculine virtues evolved … into a code of deportment founded on the basis of moral righteousness, aggressive physicality, male sexuality and camaraderie, self-control and stoic attention to duty': Stephen Heathorn, '"The highest type of Englishman": Gender, War and the Alfred the Great millenary commemoration of 1901', *Canadian Journal of History*, 37 (2002), 459–482 (p. 472).

32 John Tosh, 'Domesticity and manliness in the middle-class family of Edward White Benson', in *Manful Assertions: Masculinities in Britain since 1800*, ed. by Michael Roper and John Tosh (London: Routledge, 1991), pp. 44–73 (pp. 51–53).

33 B.J. Becker, 'Dispelling the myth of the able assistant: Margaret and William Huggins at work in the Tulse Hill Observatory', in Pycior, pp. 98–111.

34 Orton, pp. 34–35.

35 LGL, 21,907–8 (The New Electrical Laboratories', *The Central*, 1 (4) (1903), 7–12).

36 J.A. Kesiner, *Sherlock's Men: Masculinity, Conan Doyle and Cultural History* (Aldershot: Ashgate, 1997), pp. 15–18.

37 Thomas A. Sebeok and Harriet Margolis, 'Captain Nemo's Porthole: Semiotics of Windows in Sherlock Holmes', *Poetics Today*, 3 (1) (1982), 110–139 (p. 110).

38 See Shawn Rosenheim, '"The King of 'Secret Readers'": Edgar Poe, Cryptography, and the origins of the Detective Story', *English Literary History*, 56 (2) (1989), 375–400.

39 Sebeok and Margolis, p. 112.

40 Wells, who much admired T.H. Huxley, Darwin's vocal supporter, had received a scholarship to train as a science teacher. This provided free tuition and a stipend of a guinea a week. Norman and Jeanne Mackenzie, *The Life of H.G. Wells: The Time Traveller* (London: Hogarth Press, 1987), p. 53.

41 Foppish Manning, visiting the laboratory, 'carried a cane and a silk hat with a mourning-band in one grey-gloved hand; his frock coat and trousers were admirable … The low ceiling made him seem abnormally tall'. However, the heroine admired scientist Capes and 'felt him as something solid, strong and trustworthy': H.G. Wells, *Ann Veronica* (London: Everyman/J.M. Dent, 1943, repr. 1999), pp. 196–197 and p. 130.

42 See Robin Gilmour, *The Victorian Period: The Intellectual and Cultural Context of English Literature, 1830–1890* (Harlow: Longmans, 1993), p. 31.

43 'At last there seems a fair prospect that …', *The Times*, July 3 1905, editorial/leader, p. 9.

44 Mackenzie, *Life of H.G. Wells*, p. 55.

45 'Universities and the State', *Nature*, 70 (1812) July 21 1904, News, p. 271.

46 *The Times*, 'Aspects of Eugenics', July 25 1912, p. 9.

47 Gooday, 'Teaching Telegraphy', p. 99.

48 Henry E. Armstrong, 'Technical Education in Ireland', *The Times*, June 11 1901, Letters to Editor, p. 12. The dispute centred on a new scheme of technical education, like that at Finsbury, being introduced into Ireland.

49 Henry J. Spooner, 'The Education of Engineers', *The Times*, July 19 1911, Letters to Editor, p. 24.

50 LUSA, Young Papers, D140/35/20 (William Young's notes).

51 Armstrong, 'Mrs Hertha Ayrton', p. 801.

52 LGL, 21,868/10.

53 LGL, 21,907–8 ('The Mosely Commission', *Central*, 1 (4) (June 1904), 122–126). The Mosely Commission reported on education in America. Henry E. Armstrong (1848–1937), professor of chemistry, is remembered as an educational reformer who lobbied for heuristic learning.

54 For example, J.J. Thomson, who was sceptical of female intellectual capacities and denied female researchers equal access to the Cavendish Laboratory; and William Huggins and Joseph Larmor, members of the Council of the Royal Society, who were critical that commercial interests tainted scientific work and were strongly opposed to the admission of women.

55 Sharp, *Hertha Ayrton*, pp. 186–187.

56 LGL, 21,955 (T. Mather, 'William Edward Ayrton', *The Central*, 7 (21) (1910), 70–80) (Memorial issue to W.E. Ayrton).

57 Sharp, *Hertha Ayrton*, p. 150.

58 Armstrong, 'Mrs Hertha Ayrton', p. 801.

59 Research required burning carbons over long periods of time, watching for minute changes. Mather supervised the first set of experiments and designed succeeding ones which incorporated a mirror which projected the phenomenon onto a 'screen' of cartridge paper: Hertha Ayrton, *The Electric Arc* (London: The Electrician Printing and Publishing Company, 1903), p. 352.

60 IEE, 'The Arc Lamp' <http://archives.iee.org/about/Arclamps/arclamps.html> [accessed February 11 2005].

61 In 1878, the executive committee of the City and Guilds of London Institute recommended a national scheme of technical education including a central institute for advanced instruction and research in science and technology, and the development of local trade schools. The 1883–4 programme states as one of its objects the education of 'Persons of either sex who wish to receive a scientific and practical training …', LGH, 21,861 and 21,970.

62 LGL, 21,868/10.

63 Graeme J.N. Gooday, 'The premisses of premises: Spatial issues in the historical construction of laboratory credibility', in *Making Space for Science: Territorial Themes in the Shaping of Knowledge*, ed. by Crosbie Smith and John Agar (Basingstoke: Macmillan, 1998), pp. 216–245 (p. 238).

64 Ibid., p. 238.

65 For example the work of H. Rider Haggard and R.L. Stevenson. These narratives have been interpreted as character-building tales which comprise a tradition of writing important to supplying British Imperialism with an energising myth. See Martin Burgess Green, *Dreams of Adventure: Deeds of Empire* (London: Routledge and Kegan Paul, 1980), pp. 3–8.

66 LGL, 21,907–8 ('The Vacuum Cleaner', *The Central*, 1 (4) (1904), 146–148).

67 C.S. Bremner, *Education of Girls and Women in Great Britain* (London: Sonnenschein, 1897), p. 221.

68 Cynthia Scheinberg, 'Re-mapping Anglo-Jewish literary history', *Victorian Literature and Culture*, 27 (1) (1999), 115–124 (p. 118).
69 Tattersall and McMurran, p. 86.
70 Paris, Bibliothèque nationale, Papers, Naf. 18443, fol. 310 (letter, January 7 1912).
71 Autumn Stanley writes that until the late 1970s 'no book had ever been done on women as inventors. Not only were there no books, but even book chapters and articles on women as contributors to technology were vanishingly rare'. Stanley's pioneering book provides comprehensive redress: Autumn Stanley, *Mothers and Daughters of Invention: Notes for a Revised History of Technology* (New Brunswick NJ: Rutgers University Press, 1995), p. xvii.
72 Schiebinger, pp. 136–144; Fox Keller, pp. 43–65.
73 LGL, 21,9956 ('The electrical engineers', *Central Gazette*, 2 (1) (1899), 18–20). Their interests included electricity for harbour defence, mines and searchlights.
74 Warner, p. xxi.
75 Ibid., p. xx.
76 Schiebinger, pp. 136–150.
77 Ibid., p. 134.
78 IEE (Trotter, pp. 569–591).
79 H.J. Mozans, *Woman in Science, with an introductory chapter on woman's long struggle for things of the mind*, 3rd edn (Notre Dame, I: University of Notre Dame Press, 1991), p. 252. Eleanor Ormerod (1828–1901) was a self-taught investigator, author and public speaker; from 1882–1892 she was consulting entomologist to the Royal Agricultural Society, she was a fellow of the Royal Meteorological Society and an examiner in agricultural entomology at the University of Edinburgh, which in 1900 awarded her the first honorary LL.D it had ever offered to a woman. Ogilvie, *Women in Science*, pp. 142–143.
80 Mozans, p. 224.
81 Burstyn, pp. 109–110.
82 Mozans, pp. 408–410.
83 Armstrong, 'Mrs Hertha Ayrton', p. 801.
84 Ibid.
85 Martin J. Wiener, *English Culture and the Decline of the Industrial Spirit, 1850–1980* (Cambridge: Cambridge University Press, 1981), pp. 5–18.
86 Frevert, pp. 107–137.
87 The elitist, anti-state, anti-technical position of some 'feminist' but anti-suffrage opinion has been demonstrated by Lucy Delap: '"Philosophical vacuity and political ineptitude" *The Freewoman*'s critique of the suffrage movement', *Women's History Review*, 11 (4) (2002), 613–630.
88 Schiebinger, pp. 144–146.
89 Kathryn A. Neeley, *Mary Somerville: Science, Illumination and the Female Mind* (Cambridge: Cambridge University Press, 2001), p. 188.
90 Joy Harvey, '*Almost a Man of Genius': Clémence Royer, Feminism and Nineteenth-century Science* (New Brunswick, NJ: Rutgers University Press, 1997), p. 53.
91 Exeter University, Library (Special Collections), correspondence Hertha Ayrton to Norman Lockyer, January 16 1911.
92 IEE, correspondence Hertha Ayrton to Sylvanus Thompson, May 27 1899.
93 Hertha Ayrton, 'Anti-Gas Fans: Their utility in France', *The Times*, May 10 1920, Letters to Editor, p. 8.

94 Sharp, *Hertha Ayrton*, p. 246. Argon gas had been discovered by William Ramsay and Lord Rayleigh in 1894 and both were later awarded a Nobel prize. Ramsay, Professor of Chemistry at University College London, was known for his anti-feminist views.

95 Armstrong, 'Mrs Hertha Ayrton', p. 800. Armstrong cast Matilda Chaplin, Ayrton's first wife, in the role of the more gentle and 'feminine' Melisande.

96 *The Times*, 'Mrs Hertha Ayrton: A distinguished Woman Scientist', August 28 1923, Obituaries, p. 11.

Chapter 4

1 LUSA, Young Papers, D140/9/146. George M. Minchin (1845–1914) had been a Professor of Mathematics at the Royal Indian Engineering College and the University of London; was elected a fellow of the Royal Society in 1895.

2 I. Grattan-Guinness, 'A mathematical union: William Henry and Grace Chisholm Young', *Annals of Science*, 29 (2) (1972), 105–186 (pp. 140–141).

3 For example see Graham Sutton, 'The centenary of the birth of W.H. Young', *Mathematical Gazette*, 59 (1963), 17–21.

4 Sylvia Wiegand, 'Grace Chisholm Young and William Henry Young: A partnership of itinerant British mathematicians', in Pycior et al., *Creative Couples*, pp. 126–140.

5 Ibid., p. 126.

6 Ibid., p. 130 and p. 135.

7 Ibid., p. 138.

8 Ibid., p. 139 and p. 140.

9 LUSA, Tanner Papers, D599/16 and Young Papers, D140/12 (both are Grace's autobiographical notes).

10 Genevieve Lloyd, *The Man of Reason: 'Male' and 'Female' in Western Philosophy* (London: Methuen, 1984), p. 76.

11 LUSA, Young Papers, D140/6/44 (October 23 1893).

12 Fernanda Perrone, 'Women academics in England, 1870–1930', *History of Universities*, 12 (1) (1993), 339–367 (p. 339).

13 Helena Swanwick, 'Memoir of Girton, 1882–1885', in *Strong-Minded Women and Other Lost Voices from Nineteenth-century England*, ed. by Janet Murray (Harmondsworth: Penguin, 1992), pp. 239–242 (p. 240).

14 LUSA, Young Papers, D140/2/2.1 (Cecily Tanner's notes on her parents).

15 Sutton, p. 19.

16 LUSA, Young Papers, D140/8/81 (courtship correspondence, Young to Grace, October (n.d.) 1895).

17 Delamont, p. 142.

18 Barbara Caine, *Destined to be Wives: The Sisters of Beatrice Webb* (Oxford: Clarendon Press, 1986), p. 141.

19 LUSA, Young Papers, D140/6/60 (letter from Grace to Frances de Grasse Evans (n.d.) c.1893).

20 Lidia Sciama, 'Ambivalence and dedication: Academic wives in Cambridge University, 1870–1970', in *The Incorporated Wife*, ed. by Hilary Callan and Shirley Ardener (London: Croom Helm, 1984), pp. 50–66 (p. 53).

21 LUSA, Young Papers, D140/9/1 (Birkhoff to Young, August 17 1919).

22 Ibid., D140/8/201-9 (Young to Grace, April 27 1896).
23 Ibid., D140/4/6 (Young to Grace, November 1901).
24 Masculinity ... is a relational construct, incomprehensible apart from the totality of gender relations', Roper and Tosh, p. 2; 'Masculinity is never fully possessed, but must perpetually be achieved, asserted and renegotiated', Kesiner, p. 15.
25 LUSA, Young Papers, D140/6/179 (Grace to Frances de Grasse Evans, n.d., 1897).
26 Ibid., D140/6/163 (Grace to Frances de Grasse Evans, August 24 1896).
27 Ibid., D140/24/3 (Young to Grace, from Yokohama, October 1915).
28 LUSA, Tanner Papers, D599/16 (Grace notebook).
29 LUSA, Young Papers, D140/2/3 (Cecily Tanner's notes on parents).
30 Ibid., D140/6/and D/140/4 (Series of letters between Young and Grace, October 1900).
31 Pycior, p. x; Abir-am, p. 8.
32 LUSA, Young Papers, D140/6/499a (Grace to Young, November 1901. Grace writes that she will 'drag a testimonial out of Klein, however unwilling he may be').
33 Arthur Schönflies (1853–1928) had completed his doctorate with Klein at Göttingen and was then a professor of mathematics at Königsberg. He is known for his work in set theory and crystallography.
34 Brian Clegg, *Infinity: The Quest to Think the Unthinkable* (London: Robinson, 2003), p. 157. This provides an accessible history of set theory and the concept of infinity.
35 Russell's paradox is the most famous of the logical or set-theoretical paradoxes. The paradox arises within naive set theory by considering the set of all sets that are not members of themselves. Such a set appears to be a member of itself if and only if it is not a member of itself, hence the paradox: A.D. Irvine, 'Russell's Paradox', in *The Stanford Encyclopedia of Philosophy* (Summer 2004 Edition), ed. by Edward N. Zalta <http://plato.stanford.edu/archives/sum 2004/ entries/russell-paradox/> [accessed February 12 2005].
36 LUSA, Young Papers, D140/4/1 (Young to Grace, September 22 1901). Between 1909 and 1911 William Young carried out a long-running dispute with Schönflies on the pages of the *Messenger of Mathematics*.
37 LUSA, Young Papers, D140/30/1-2 (n.d., from Youngs' correspondence on mathematics, 1891–1914).
38 Ibid., D140/6/357 (Grace to Young, October 22 1900).
39 Ibid., D140/30/1 (Young to Grace, February 1904): W.H. Young, 'On upper and lower integration', *Proceedings of the London Mathematical Society*, 2 (1904–5), 52–66.
40 LUSA, Young Papers, D140/24/2 (Grace to Young, March 5 1914).
41 Ibid., D140/7/1.20 (Grace to Young, January 1906).
42 Ibid., D140/5/1 (Grace's diary 1908–1939).
43 Ibid., D140/9/47 and 49 (Hardy to Young, n.d. c.1917/18). G.H. Hardy (1877–1947) was an analyst and professor of mathematics at Cambridge, in 1919 he left that University to take up the Savilian Chair at Oxford . Young's lack of brevity in his papers was exacerbated by a paper shortage that became acute during World War One.
44 Ibid., D140/24/2 (Grace to Young, February 27 1914).

45 LUSA, Young Papers, D140/6/228 (Grace to mother, June 1898). This is quoted in Grattan-Guinness, 'Mathematical union', p. 136.
46 Ibid., D140/6/224 (Grace to Frances de Grasse Evans, June (n.d.) 1898).
47 For example: Margaret W. Rossiter, 'The Matthew/Matilda effect in science', *Social Studies of Science*, 23 (1993), 325–341; Sime, pp. 326–329, documents how Lise Meitner was overlooked for a Nobel Prize for her contribution to the discovery of nuclear fission; the prize went to her male partner Otto Hahn despite Meitner being the lead investigator.
48 Robin J. Wilson, 'Hardy and Littlewood', in *Cambridge Scientific Minds*, ed. by Peter Harmant and Simon Mitton (Cambridge: Cambridge University Press, 2002), pp. 202–219 (p. 202).
49 John J. O'Connor and Edmund F. Robertson, 'G.H. Hardy', in *The Mac-Tutor History of Mathematics Archive* <http://www-history.mcs.st-andrews.ac.uk/history/index.html> [accessed February 5 2005].
50 Wilson, 'Hardy and Littlewood', pp. 202–205.
51 Grace published one paper in 1912.
52 LUSA, Young Papers, D140/6/553 (Young to Grace, February 15 1902).
53 Women's rights campaigners in England had been working since the mid-nineteenth century to secure rights for married women, in 1882 their lobbying brought some success with the 'Married Woman's Property' Act which allowed women to retain their own property on marriage, but their person was still included within that of their husbands in other areas and, of course, women did not have the vote.
54 W.H. Young and Grace Chisholm Young, p. vi.
55 Patricia Rothman recounts an amusing anecdote (received from a friend of the Youngs, A.S. Besicovitch) which illustrates Young's insecurity: 'William Henry Young was out swimming one day with Besicovitch and he got into difficulties. Besicovitch swam over to help him. With Besicovitch's assistance, W.H. Young came up for a "third time" coughing, his long beard bobbing in the waves, he spluttered out as he gasped for breath "Are you one of those people who think my wife is a better mathematician than I am?"', Patricia Rothman, 'Grace Chisholm Young and the division of the laurels', *Notes and Records of the Royal Society of London*, 50 (1) (1996), 89–100 (p. 97).
56 Sutton, p. 21.
57 LUSA, Young Papers, D140/30/5.1 (G.H. Hardy, 'W.H. Young', *Journal of the London Mathematical Society*, 17 (1942), 218–237 (p. 220).
58 Abir-am, pp. 3–4.
59 Wiegand, 'Partnership of itinerant British mathematicians', p. 140.
60 Carol Dyhouse, *Girls Growing Up in Late Victorian and Edwardian England* (London: Routledge and Kegan Paul, 1981), p. 172.
61 Deirdre David, '"Art's a service": Social wound, sexual politics and Aurora Leigh', in *Victorian Woman Poets: Emily Brontë, Elizabeth Barrett Browning, Christina Rossetti*, ed. by Joseph Bristow (Basingstoke: Macmillan, 1995), pp. 108–131 (p. 129).
62 For a discussion of the work of L.T. Meade see Garriock, pp. 196–255; for a discussion of 'The Girton Girl: Social images from within and without', see Bradbrook, pp. 91–112.
63 LUSA, Young Papers, D140/12/22 and 23. Grace was described by May Winston as 'tall and large with red cheeks, a rather muddy complexion, and

a very pleasant voice like all the English women of education whom I have met', Winston Papers, unpublished memories, pp. 16–17.

64 LUSA, Tanner Papers, D599/16 (Grace's notebook, entry dated '25.V.17').
65 LUSA, Young Papers, D140/6/328 (Grace to Frances de Grasse Evans, December 30 1899).
66 Ibid., D140/6/329 (Frances de Grasse Evans to Grace, January 7 1900).
67 Ibid., D140/9 (Young to P. Dienes (n.d.) 1921).
68 Tosh, pp. 44–73.
69 LUSA, Young Papers, D140/24/2 (Young to Grace, January 16 1901: 'I will try and fill a father's as well as a husband's place to you').
70 Neeley, pp. 206–214.
71 Sidgwick, 'Mrs Henry Sidgwick', p. 66.
72 From *Lady's Realm* 1 (1897), 76–81, quoted in Margaret Beetham, *A Magazine of her own? Domesticity and Desire in the Women's Magazine* (London: Routledge, 1996), p. 133.
73 Becker, pp. 98–111.
74 Caine, pp. 181–197.
75 Burstall, *Retrospect and Prospect*, p. 70.
76 'Feminist' would not, however, have been a term that Grace would have been likely to apply to herself. Its origin is commonly attributed to early-nineteenth century France and the earliest reference found to its usage in English is in an article in the *Westminster Review* in 1898. See Delap, 'Philosophical vacuity and political ineptitude', p. 626.
77 Angelique Richardson, *Love and Eugenics in the Late Nineteenth Century: Rational Reproduction and the New Woman* (Oxford: Oxford University Press, 2003), p. 8. See also Sally Ledger, 'The New Woman and the crisis of Victorianism', in Ledger and McCracken, pp. 22–44.
78 Sarah Grand (1854–1943) combined commitment to women's emancipation with a belief in biological determinism and eugenics. While at Girton, Grace was influenced by Grand's 1893 book *The Heavenly Twins* in which a 'highly-bred' woman is married to a man of loose morals; the novel's themes are venereal disease, social purity and the consequent withholding of sexual favours by women.
79 All quotes in this paragraph from LUSA, Young Papers, D140/37 (misc. notes).

Chapter 5

1 Alexander Wood, *The Cavendish Laboratory* (Cambridge: Cambridge University Press, 1946), p. 4. One of Maxwell's projects was the publication of Henry Cavendish's electrical papers; connected to this, and also as a tribute to the generosity of the Cavendish family in endowing the laboratory, Maxwell deployed the resources of the Cavendish to repeat Cavendish's experiments and check his measures. To do so, Maxwell undertook investigations into the physiological effects of electric currents with himself as subject. Peter M. Harman, *The Scientific Letters and Papers of James Clerk Maxwell*, 3 vols (Cambridge: Cambridge University Press, 1990–2003), vol. 3, pp. 11–13.

2 Maxwell was not wholly antipathetic to women in science however; correspondence with his wife, Katherine Mary Clerk Maxwell, reveals com-

munication about his work and other scientific issues. Ibid., vol. 2, p. 122, p. 528, p. 623 and p. 629; vol. 3, p. 873.

3 Lord Rayleigh, Robert John Strutt, 'Some reminiscences of scientific workers of the past generation, and their surroundings', *Proceedings of the Physical Society*, 48 (2) (1936), 216–246 (p. 230).

4 More recently, 'manly' behaviour and a marginalisation of women has been shown to be a dominating factor in mid-late twentieth century particle physics laboratories in the USA and Japan. See Traweek, Chapter 3, 'Pilgrim's Progress: Male tales told during a life in physics', pp. 74–105.

5 For example see Abir-am, pp. 3–4.

6 For example, Gould, 'Women and the culture of university physics', and Richmond, 'A lab of one's own'.

7 For example, representations of biological laboratories are often informed by aesthetic considerations and refer back to different traditions. For H.G. Wells' Ann Veronica, the biological laboratory at the Central Imperial College, where she was studying comparative anatomy, 'had an atmosphere that was all its own ... (and) ... it made every other atmosphere she knew seem discursive and confused ... the room was more simply concentrated in aim than a church ... this long, quiet, methodological chamber shone like a star seen through clouds'. Wells, p. 115.

8 *Nature*, 61 (January 23 1890), News, p. 279.

9 Gould, 'Culture of university physics', pp. 132–137.

10 LGU, 29,973.

11 Bremner, p. 181.

12 It is noticeable in these adventure novels that women tend to be absent or represented as unnatural beings. As Quatermain explains prior to narrating his story: 'There is Gagoola, if she was a woman and not a fiend. But she was a hundred at least, and therefore not marriageable, so I don't count her. At any rate, I can safely say that there is not a *petticoat* in the whole history'. H. Rider Haggard, *King Soloman's Mines* (Hertfordshire: Wordsworth Editions, 1998), p. 10. (This novel was first published in 1886.)

13 Rayleigh, *Life of J.J. Thomson*, p. 51 and p. 26.

14 Ibid., p. 31.

15 Graeme J.N. Gooday, 'Ayrton, William Edward (1847–1908)', Oxford Dictionary of National Biography (Oxford: Oxford University Press, 2004) <http://www. oxforddnb.com/view/article/30509> [accessed 29 January 2005].

16 Rayleigh, 'Reminiscences', p. 226.

17 Peter Broks, *Media Science Before the Great War* (Basingstoke: Macmillan, 1996), pp. 30–31.

18 Broks, p. 43.

19 Ibid., pp. 41–51.

20 Richmond, 'A lab of one's own', p. 256.

21 Carol Dyhouse, *No Distinction of Sex? Women in British Universities, 1870–1939* (London: UCL Press, 1995), p. 144. William Ramsay was Professor of Chemistry at University College London from 1887–1913; he received a Nobel prize in conjunction with Lord Rayleigh in 1904.

22 Rayleigh, *Life of J.J. Thomson*, p. 46.

23 Gould, 'Culture of physics', p. 129. Despite this, Gould emphasises integration and argues that the women researchers did become part of a team and not interlopers.

24 Gould divides the female researchers at the Cavendish into two groups, those who were from a 'free-thinking' or foreign background and others whose presence at the Cavendish was part of a pattern of intellectual life common to Victorian gentlewomen and Cambridge families. Ibid., pp. 134–137.

25 Rayleigh, *Life of J.J. Thomson*, p. 34.

26 For example, Armstrong's obituary in *Nature* is dismissive of Hertha's scientific abilities as 'overpainted' and focusses more on personal details: Armstrong, 'Mrs Hertha Ayrton', pp. 800–801 (p. 801).

27 'Anniversary Address', *Proceedings of the Royal Society of London*, Section A, 79 (1906), 1–12 (p. 12).

28 Ibid.

29 LGL, 29,956 (*Central Gazette*, 1 (1) (1899), 75).

30 Garriock, p. 109.

31 Dyhouse, *No Distinction of Sex?*, p. 33.

32 Hertha Ayrton, *The Electric Arc*, p. vi.

33 Mather, 'Mrs Hertha Ayrton'.

34 Hertha Ayrton, 'The reason for the hissing of the electric arc', *Nature*, 60 (July 27 1899), 302–305 (p. 304).

35 Sir William Huggins, *The Royal Society, or, Science in the State and in the Schools* (London: Methuen, 1906), p. 40.

36 Royal Society of London, 'Sir George Gabriel Stokes', in *Proceedings of the Royal Society of London: Containing Obituaries of Deceased Fellows, Chiefly for the Period 1898–1904*, 75 (1905), 210–216 (p. 215).

37 Larry Owens, 'Pure and sound government: Laboratories, playing fields and gymnasia in the nineteenth-century search for order', *ISIS* 76 (1985), 182–194 (p. 193). (Owen's focus is on the USA.)

38 Kesiner, p. 2.

39 Rayleigh, *Life of J.J. Thomson*, p. 15.

40 Lord Rayleigh, 'Address of the president, Lord Rayleigh, O.M., D.C.L., at the anniversary meeting on November 30th, 1907', *Proceedings of the Royal Society of London, Series A*, 80 (1907–8), 231–251 (p. 243).

41 Felix Klein and the Göttingen School of Mathematics attracted similar suspicions in the era before World War Two, articulated by way of a supposed conspiracy of Jewish mathematicians: 'Jews were innately inclined toward algorithmic, analytic, or abstract thinking, whereas Germans tended to think intuitively and synthetically, often drawing their inspiration from natural phenomena.' David E. Rowe, '"Jewish Mathematics" at Göttingen in the era of Felix Klein', *ISIS*, 77 (3) (1986), 422–449 (p. 424).

42 There was contention surrounding Marie's nomination: she was deliberately excluded from the first nomination, her name only included after pressure from an influential member of the Swedish Academy of Sciences, mathematician Gustav Mittag-Leffler. See Susan Quinn, *Marie Curie: A Life* (Cambridge MA: Da Capo Press, 1995), pp. 187–190.

43 Helena M. Pycior, 'Pierre Curie and "His eminent collaborator Mme Curie": Complementary partners', in Pycior, pp. 39–56 (p. 56).

44 Shirley Ardener, 'Introduction', in *Women and Space: Ground Rules and Social Maps*, 2nd edn, ed. by Shirley Ardener (Oxford: Berg, 1993), pp. 1–15 (p. 5).

45 Simon Schaffer, 'Physics laboratories and the Victorian country house', in *Making Space for Science: Territorial Themes in the Shaping of Knowledge*, ed. by Crosbie Smith and John Agar (Basingstoke: Macmillan, 1998), pp. 149–180 (p. 177).

46 A.T. Humphrey, 'Lord Rayleigh: The last of the great Victorian polymaths', *GEC Review*, 7 (3) (1992), 167–180. Available online: <http://www.marconi. com/Home/about_us/Our%20History/Publications%20Archive> [accessed February 13 2005].

47 Rayleigh, 'Reminiscences', pp. 238–240.

48 Broks, p. 37.

49 For discussion of the expansion in scale, cost and time of experiments from this period see Nicholas Jardine, *The Scenes of Inquiry: On the Reality of Questions in the Sciences* (Oxford: Clarendon Press, 1991), pp. 94–120.

50 See Graeme J.N. Gooday, *The Morals of Measurement: Accuracy, Irony, and Trust in Late Victorian Electrical Practice* (Cambridge: Cambridge University Press, 2004), pp. 23–39.

51 Schaffer, pp. 151–152.

52 Gooday, 'Premises of premises', pp. 219–222.

53 For a description of the work of Pearson's eugenics laboratory, see M. Eileen Magnello, 'The non-correlation of biometrics and eugenics: Rival forms of laboratory work in Karl Pearson's career at University College London, Part 1', *History of Science*, 37 (1) (1999), 79–106 and 'Part 2', *History of Science*, 37 (2) (1999), 123–150, especially Part 2, pp. 126–136.

54 Concern was raised that the pressure indicators attached to the tanks of water that she had used could have leaked and distorted the readings. Hertha redesigned the apparatus and her paper was finally accepted.

55 Royal Society of London, Library, Archives and Manuscripts (RSL), Referee Report 143/1904 (The author was John Joly, Professor of Geology and Mineralogy at the University of Dublin).

56 Hertha Ayrton, 'The origin and growth of Ripple Mark', *Proceedings of the Royal Society of London*, 84 (1910), 285–310.

57 Tattersall and McMurran, p. 103.

58 IEE, Trotter Memoirs, p. 587.

59 Sharp, *Hertha Ayrton*, p. 282.

60 Royal Society, 'Sir George Gabriel Stokes', p. 211; Humphrey, 'Lord Rayleigh', p. 8.

61 Cargill G. Knott, ed., *Collected Scientific Papers of John Aitken: Edited for the Royal Society of Edinburgh, with an Introductory Memoir by Cargill G. Knott* (Cambridge: Cambridge University Press, 1923), pp. vii–xii. Aitken (1839–1919) researched air formations and other atmospheric phenomena.

62 Thanks to the graduate seminar in history and philosophy of science, University of Leeds 2003, for this insight.

63 Shapin and Schaffer, especially Chapter 4: 'The trouble with experiment': Hobbes versus Boyle', pp. 110–154.

64 Porter, especially Chapter 9: 'Is science made by communities?', pp. 217–225.

65 Trotter, *Memoirs*, p. 587.

66 Royal Holloway, University of London, Archive and Records, Bedford College Magazine, AS200/3/61 ('Mrs Ayrton and the Hughes Medal', *Bedford College Magazine*, 61 (December 1906), 13–14 (p. 14)).

67 Sharp, *Hertha Ayrton*, p. 156. Male scientists were often positioned within the narrative of eccentric genius, but for a woman to be so is more problematic as there is no tradition of such representation; the feeling is of amusement but without the link to 'greatness'.
68 Zangwill, p. 9.
69 Ibid., p. 131.
70 Wells, p. 196. There is a further interesting subtext to this as the (unsuccessful) suitor is constructed as unmasculine and foppish because he is not at ease in scientific surroundings, as discussed in Chapter 3.
71 Sharp, *Hertha Ayrton*, p. 154.
72 Tattersall and McMurran, p. 88.
73 Margaret W. Rossiter, '"Women's Work" in Science', in Kohlstedt, pp. 287–304 (p. 289).
74 National Physical Laboratory, '100 years of the National Physical Laboratory', *Metromnia*, 9 (2000) <http://www.npl.co.uk/npl/publications/metromnia/> [accessed February 5 2005].
75 See Graeme Gooday, 'Precision measurement and the genesis of physics teaching laboratories in Victorian Britain', *British Journal for the History of Science*, 23 (1990), 25–51 (pp. 29–36).
76 Isabel Falconer, 'J.J. Thomson and "Cavendish Physics"', in James, *The Development of the Laboratory*, 104–117 (p. 106).
77 Humphrey, 'Lord Rayleigh', p. 10.
78 Broks, p. 44.
79 Lord Rayleigh, 'Presidential address', p. 240; P. Zeeman, 'Scientific worthies: Sir William Crookes, F.R.S., *Nature*, 77 (November 7 1907), 1–3 (p. 1).
80 Rayleigh, 'Reminiscences', p. 237.

Chapter 6

1 Statistics of women mathematicians at British Universities from *Davis Archive of Female Mathematicians* <http://www-history.mcs.st-andrews.ac.uk/history/Davis/info.html> [accessed March 25 2009].
2 See Forfar, 'What became of the Senior Wranglers?'.
3 Dyhouse, *No Distinction of Sex?*, p. 134.
4 For an analysis of relations between the Cambridge mathematics and natural sciences triposes see Wilson, 'Experimentalists among the mathematicians'. Wilson omits women from his analysis.
5 J.W.L. Glaisher, 'Presidential address', p. 29. Glaisher (1848–1928) was a lecturer at Trinity College Cambridge for most of his adult life and a Fellow of the Royal Society. In 1896, Glaisher was reported to be the only mathematician at Cambridge who closed his lectures to women.
6 'The mathematics tripos', *Girton Review* (December 1882), 3–4.
7 Tripos results are reproduced in J.R. Tanner, ed., *The Historical Register of the University of Cambridge, being a supplement to the Calendar with a record of university offices, honours and distinctions to the year 1910* (Cambridge: Cambridge University Press, 1917).
8 Warwick, *Masters of Theory*.
9 Barrow-Green, p. 271.

10 Ibid., p. 303. Barrow-Green links her statistics to research interests of Cambridge's leading mathematicians at the time.
11 See Richmond, 'A lab of one's own'; and Marsha L. Richmond, 'Women in the early history of genetics: William Bateson and the Newnham College mendelians, 1900–1910', *ISIS*, 92 (2001), 55–90.
12 Vickery, p. 155. Newnham's Ida Freund was a chemist, teacher, and researcher at the Cavendish Laboratory.
13 Dyhouse, *No Distinction of Sex?*, p. 46.
14 Rothblatt, pp. 210–242.
15 Howson, p. 177.
16 *Bedford College for Women Calendar*, 1910–11.
17 Burstall, 'Place of mathematics in girls' education'.
18 Mary T. Brück, 'Alice Everett and Annie Russell Maunder, torch bearing women astronomers', *Irish Astronomical Journal*, 21 (3/4) (1994), 281–291 (p. 282).
19 Alice Everett (1865–1949) moved to the Astrophysical Observatory at Potsdam in 1896 after five years at Greenwich and later worked at the Vassar College Observatory, USA, with Maria Mitchell, Professor of Astronomy. Annie Russell Maunder (1868–1947) married Edward Walter Maunder, her supervisor in the Solar Department at Greenwich, in 1895 and resigned her post. She continued to be actively involved in astronomy as editor of the British Astronomical Association Journal and in collaboration with her husband on astronomical expeditions and writing books. In 1916, when the Royal Astronomical Society opened its doors to women, she became one of its first female fellows.
20 Brück, p. 282. Caroline Herschel (1750–1848) discovered eight comets and worked with her brother William. She was granted a salary of £50 per year by King George III.
21 Peggy Aldrich Kidwell, 'Women astronomers in Britain, 1780–1930', *ISIS*, 75 (3) (1984), 534–546.
22 David Alan Grier, *When Computers were Human* (Princeton: Princeton University Press, 2005), p. 106.
23 Other women included Cecily Fawcett, Mary Beeton and Martha Whiteley. See Magnello. 'Part 1', pp. 87–90; Rosaleen Love, 'Alice in Eugenics-Land: Feminism and eugenics in the scientific careers of Alice Lee and Ethel Elderton', *Annals of Science*, 36 (2) (1979), 145–158; Greta Jones, 'Bell, Julia (1879–1979), geneticist', *Oxford Dictionary of National Biography* (Oxford: Oxford University Press, 2004) <http://oxforddnb.com/> [accessed September 18 2007].
24 Glaisher, p. 725.
25 'Margaret Theodora Meyer', *Girton Review*, May Term (1924); Girton College, 'Meyer, Margaret Theodora', in *Girton College Register, 1869–1946*, ed. by K.T. Butler and H.I. McMorran (Cambridge: Privately printed, 1948), p. 636.
26 Alicia Boole Stott was largely self-taught and publication of her mathematical papers resulted from her correspondence with a mathematics professor in the Netherlands who shared her interests. She was the daughter of George Boole, Professor of Mathematics at Queens College Cork. H.S.M. Coxeter, 'Alicia Boole Stott (1860–1940)', in *Women in Mathematics*, ed. by Lynn M. Osen (Cambridge MA: MIT Press, 1974), pp. 220–225.
27 Kidwell, p. 538.

28 Samuel Roberts, 'Presidential address', *Proceedings of the London Mathematical Society*, 14 (1882–3), 6–12 (p. 6).

29 Mary Lucy Cartwright, 1900–1998, gained her doctorate at Oxford and was appointed lecturer at Girton in 1935, becoming Mistress 1946–1960. In 1947 she became the first female mathematician to be elected to the Royal Society; she won the Royal Society Sylvester Medal in 1964 and the LMS De Morgan medal in 1968. Her subject was the theory of functions, her work making a contribution to the development of chaos theory. Shawnee L. McMurran and James J. Tattersall, 'Mary Cartwright, 1900–1998', *Notices of the American Mathematical Society*, 46 (2) (1999), 214–220.

30 Hudson also spent 1912–1913 in Bryn Mawr with Professor of Mathematics there Charlotte Angas Scott. At the outbreak of World War One she joined the Civil Service to undertake work for the Air Ministry on applied probability problems associated with aircraft. Newnham College, 'Hilda Phoebe Hudson, 1881–1965', *Newnham College Roll Letter* (1966), 53–54.

31 R.A. Rankin, 'The first hundred years (1883–1983)', *Proceedings of the Edinburgh Mathematical Society*, 26 (1983), 135–150, p. 137.

32 Flora Philip (1865–1943) was assistant mistress of St. George's High School for Girls from 1888. In 1893 she was one of the first group of women to be awarded degrees from the University of Edinburgh; she married shortly after and gave up teaching and EMS membership. See http://www-groups.dcs.st-and.ac.uk/~history/Printonly/Philip_Flora.html [accessed March 25 2009].

33 Jessie Chrystal Macmillan (1872–1937) gained a first-class honours degree in mathematics and natural philosophy in 1896, and an MA in mental and moral philosophy in 1900, both awarded by the University of Edinburgh. She was active in the women's suffrage campaign in Scotland and London and, in 1924, became one of the first generation of women barristers. Sybil Oldfield, 'Macmillan, (Jessie) Chrystal (1872–1937)', *Oxford Dictionary of National Biography*, Oxford University Press, 2004 <http://www.oxforddnb.com/view/article/38526> [accessed 27 March 2009].

34 Elizabeth A. McHarg (1923–1999). McHarg graduated from the University of Glasgow in 1943 and, as a research student at Girton College, Cambridge, studied under Mary Cartwright. She then returned to Glasgow as a lecturer in mathematics for the next forty years. Dan Martin, 'Elizabeth A. McHarg', *Glasgow Mathematical Journal*, 42 (2000), 487–488.

35 Della Dumbaugh Fenster and Karen Hunger Parshall, 'Women in the American Mathematical Research Community: 1891–1906', in *The History of Modern Mathematics*, vol. III, *Images, Ideas and Communities*, ed. by E. Knobloch and D. Rowe (San Diego: Academic Press, 1994), p. 229.

36 Judy Green and Jeanne LaDuke, 'Women in American mathematics: A century of contributions', in *A Century of Mathematics in America, Part II*, ed. by Peter Duren with the assistance of Richard A. Askey and Uta C. Merzbach (American Mathematical Society: 1989), pp. 379–398, p. 384.

37 Fenster and Parshall, pp. 229–230. Ida May Schottenfels (1869–1942) was awarded a masters degree in mathematics from the University of Chicago in 1896 but never went on to gain a doctorate; she was a teacher of mathematics in high schools for most of her life.

38 Green and LaDuke, *Century of Contributions*, p. 384.

39 Augustus De Morgan, 'Presidential address', *Proceedings of the London Mathematical Society*, 1 (1865–1866), 1–4.

40 A. Rupert Hall, *The Cambridge Philosophical Society: A History, 1819–1969* (Cambridge: Cambridge Philosophical Society, 1969), p. 66.

41 Ibid., p. 66.

42 Jim Tattersall and Shawnee McMurran, 'Women and the educational times' <http://www.math.csusb.edu/faculty/mcmurran/hpm.doc> [accessed February 5 2005].

43 Memberships lists of the LMS reveal that in 1905 fourteen women comprised 5.5% of membership; this is a higher percentage than is the case for 1970 when just 5% (48) of members were women. Claire Jones, 'Grace Chisholm Young: Gender and mathematics around 1900', *Women's History Review*, 9 (4) (2000), 675–692 (p. 679). Similar trends have been identified elsewhere. Patricia Kenschaft writes of the AMS 'It is sobering to compare the measurable progress of women in academia during the 1970s with that round the turn of this (twentieth) century ... it appears that in the mathematical community, women have merely regained their former position'. Patricia C. Kenschaft, 'Women in mathematics around 1900', *Signs*, 7 (4) (1982), 906–909, p. 906. Her opinion is echoed by Green and LaDuke who found that the 14% of PhDs awarded in the US to women before 1940 was only matched again in the 1980s; see Judy Green and Jeanne LaDuke, *Pioneering Women in American Mathematics: The Pre-1940 PhDs* (American Mathematical Society, 2009). In Germany, Renate Tobies has found similar evidence and remarks that lower levels of female participation today 'make it difficult to imagine that mathematics could once have been one of the most popular courses of studies for women'. Tobies, p. 25.

44 Green and LaDuke, *Century of Contributions*, p. 382. Achsah Ely (1845–1904) graduated from Vassar in 1868 and was later appointed Professor of Mathematics in 1887, a post she held until the end of her life; Charlotte Cynthia Barnum graduated from Vassar in 1881, after a time teaching she resumed study at Johns Hopkins 1890–1892, after which she moved to Yale and became in 1895 one of the first three women to receive PhDs in mathematics there before 1900.

45 Winston papers, personal recollections, p. 1 and correspondence October 15 1893.

46 Fulvia Furinghetti, 'The emergence of women on the international stage of mathematics education', *ZDM Mathematics Education*, 40 (529) (2008), 529–543, p. 533. Massarini graduated from the University of Naples in 1887.

47 Donald J. Albers, G.L. Alexanderson, and Constance Reid, *International Mathematical Congresses: An Illustrated History 1893–1986* (Berlin: Springer-Verlag, 1987), pp. 6–7.

48 *New York Times*, May 5, 1935. Einstein wrote 'In the judgment of the most competent living mathematicians, Fräulein Noether was the most significant creative mathematical genius thus far produced since the higher education of women began.' http://www-history.mcs.st-andrews.ac.uk/Obits2/Noether_ Emmy_ Einstein.html [accessed September 20 2007].

49 Furinghetti, pp. 533–534.

50 Miss H.P. Hudson, 'On binodes and modal curves'. See E.W. Hobson and A.E.H. Love, eds, *Proceedings of the Fifth International Congress of Mathematicians ... 1912* (2 vols) (Cambridge: Cambridge University Press, 1913).

51 E.B. Elliott, 'Some secondary needs and opportunities of English mathematicians', *Proceedings of the London Mathematical Society*, 30 (1898–1899), 5–23 (p. 15).
52 Orton, p. 198.
53 Tobies, p. 33.
54 Green and LaDuke, *Pre-1940 PhDs*, p. 20.
55 Heathorn, pp. 467–468.
56 Warwick, *Masters of Theory*, p. 224.
57 Orton, p. 39.
58 Glaisher, p. 725.
59 For contextualisation of the relations between pure mathematics and mathematical physics in the nineteenth century see John Heard, 'Of apes and algebraists', *Viewpoint: Newsletter of the British Association for the History of Science*, 83 (June 2007), 1–3.
60 Warwick, *Masters of Theory*, effectively demystifies this mythology of genius by illustrating how Cambridge's mathematical training imparted mathematical skills and successfully reproduced its mathematical elite.

Chapter 7

1 See Tattersall and McMurran, p. 101 and p. 103; and Ogilvie, p. 33.
2 Sharp, *Hertha Ayrton*, pp. 150–170.
3 Ibid., Preface.
4 Barbara Ayrton Gould (1888–1950) was an active suffrage campaigner who became a member of the National Executive of the Labour Party in 1930; in 1945 she was elected as Member of Parliament for North Hendon.
5 Sharp advertised on the letters pages of *The Times* for any readers with information about Hertha to come forward: Evelyn Sharp, 'The late Mrs Hertha Ayrton', *The Times*, March 7, 1925, Letters to Editor, p. 8.
6 See Angela V. John, '"Behind the locked door": Evelyn Sharp, suffragette and rebel journalist', *Women's History Review*, 12 (1) (2003), 5–13.
7 Women over thirty years of age won the vote in 1918, at a time when the age bar for men was twenty-one. In 1928 women gained the vote on the same terms as men.
8 Ogilvie, p. 163.
9 Dwight Atkinson, *Scientific Discourse in Sociohistorical Context: The Philosophical Transactions of the Royal Society of London, 1675–1975* (London: L. Erlbaum, 1999), p. 102.
10 Schiebinger, p. 263. German-born Caroline Herschel (1740–1848) discovered eight comets in the closing years of the eighteenth century and her reworking of John Flamsteed's observations, *Catalogue of Stars*, was published by the Royal Society in 1798.
11 Dorothea Bate (geologist); Florence M. Durham (geneticist); Alice M. Waller; Frances Cave-Brown-Cave (mathematician); and Hertha Ayrton.
12 For a comprehensive list, see Creese, *Ladies in the Laboratory?*
13 Joseph Larmor (1857–1942) was lecturer and (from 1903) Lucasian Professor of Mathematics at Cambridge.
14 RSL, Letter 04029 (Hertha Ayrton to Joseph Larmor, Royal Society, June 12 1904).

15 Mrs. Hertha Ayrton, 'The origin and growth of Ripple Mark', *Proceedings of the Royal Society of London, Series A*, 74 (1905), abstract, pp. 565–566. Communicated by Prof. W.E. Ayrton, FRS, Received April 21; Revised May 26, Read June 16 1904.

16 Mrs. Hertha Ayrton, 'The origin and growth of Ripple-Mark', *Proceedings of the Royal Society of London, Series A*, 84 (1910), 285–310.

17 RSL, Scrapbooks (*The Times*, May 10 and June 22 1900).

18 Mrs. Hendrina (Rina) Victoria Scott, née Klaassen (d. 1929), studied Botany at the Royal College of Science with Dr. D.H. Scott; they later married and collaborated in scientific research. Scott also collaborated with botanist Ethel Sargant, as well as researching and publishing several papers in her name alone on fossil botany. Scott was one of the earliest female fellows of the Linnaean Society (elected 1905). A pioneer of the use of photography in research into the growth of plants, Scott exhibited her work at the Royal Horticultural and Linnaean Societies as well as the Royal Society and is also remembered as an early woman film-maker. 'Mrs Hendrina (Rina) Victoria Scott', *Proceedings of the Linnaean Society* (1928–29), 146–147 (Obituaries).

19 Dorothea Bate (1878–1951) was a palaeontologist and fossil hunter who had a long association with the British Museum; she was awarded a grant by the Royal Society to excavate in Cyprus, research she carried out independently. Karolyn Shindler, *Discovering Dorothea: The Life of the Pioneering Fossil-Hunter Dorothea Bate* (London: HarperCollins, 2005).

20 Gertrude Elles (1872–1960) Geologist and palaeontologist, Elles took the natural sciences tripos at Newnham College Cambridge in 1894 and was awarded a DSc by Trinity College Dublin in 1907. She taught at Cambridge and is remembered for her significant contributions to British stratigraphic geology and evolutionary research. Mary R.S. Creese, 'Elles, Gertrude Lilian (1872–1960)', *Oxford DNB*.

21 Edith Saunders (1865–1945) Botanist who sat the natural sciences tripos at Newnham College Cambridge in 1887/88 winning first-class honours and a scholarship. She later took up a teaching post at Newnham, collaborated with William Bateson on plant genetics and became Director of the Balfour Laboratory in 1899. Mary R.S. Creese, 'Saunders, Edith Rebecca (1865–1945)', *Oxford DNB*.

22 The Royal Society's 2008 Fellowship figures reveal that out of 1302 Fellows, 67 (5%) are women; however in the previous seven years 10% of new Fellows have been women (33 of 344).

23 Marion Farquharson (1846–1912) published *A Pocket Guide to English Ferns* in 1881.

24 See Joan Mason, 'The women fellows' jubilee', *Notes and Records of the Royal Society of London*, 49 (1) (1995), 125–140, p. 126.

25 Jonathon Rose, *The Edwardian Temperament, 1895–1919* (Ohio: Ohio University Press, 1986), p. 120.

26 Huggins, p. 40.

27 Mason, 'Hertha Ayrton and the admission of women', p. 207.

28 RSL, Letter 948 (William Huggins to Joseph Larmor, November 2 1906).

29 Becker, 'Able assistant'.

30 Crosbie W. Smith and Norton M. Wise, *Energy and Empire: A Biographical Study of Lord Kelvin* (Cambridge: Cambridge University Press, 1989), p. xix.

31   Brian Wynne, 'Natural knowledge and social context: Cambridge physicists an the luminous ether', in Barnes and Edge, pp. 212–231 (p. 217).

32   LUSA, Young Papers, D140/35/11 and D140/35/18 (mathematical notes and autobiographical notes).

33   Crystallographer Kathleen Lonsdale and microbiologist Marjorie Stevenson.

34   RSL, Scrapbooks (*World Magazine*, September 17 1901).

35   Eagleton argues that Victorian reason started to crumble in the 1890s as the 'high-rationalist subject of Mill or Middlemarch imploded into Madame Blavatsky and Dorian Gray': Terry Eagleton, 'The flight to the real', in Ledger and McCracken, pp. 11–21 (p. 11).

36   Helen Kanitkar, '"Real true boys": Moulding the cadets of Imperialism', in *Dislocating Masculinity: Comparative ethnographies*, ed. by Andrea Cornwell and Nancy Lindisfarne (London: Routledge, 1994), pp. 184–196 (p. 185).

37   Hamerton, pp. 263–264.

38   Quoted in Sally Ledger, *The New Woman: Fiction and Feminism at the Fin de Siècle* (Manchester: Manchester University Press, 1997), p. 117.

39   Abbott, p. 50.

40   Sir Almroth Wright, 'Militant hysteria', *The Times*, March 28 1912, Letters to Editor.

41   The classic statement of this position is Carolyn Merchant, 'Isis' consciousness raised', in Kohlstedt, pp. 11–22; also Fox Keller, pp. 7–8 and 78–80; Ludmilla Jordanova, *Sexual Visions: Images of Gender in Science and Medicine between the Eighteenth and Twentieth Centuries* (Wisconsin: University of Wisconsin Press, 1993), pp. 87–110.

42   Fox Keller, pp. 43–65.

43   Huggins, p. ix.

44   Royal Society, *The Celebration of the 250th Anniversary of the Royal Society, July 15–19, 1912* (London: Royal Society, 1913), p. 120.

45   See Schiebinger, pp. 119–159.

46   Stefan-Ludwig Hoffmann, 'Civility, male friendship and Masonic sociability in nineteenth-century Germany', *Gender and History*, 13 (2) (2001), 224–248 (pp. 236–237).

47   Royal Society, pp. 5–8.

48   When attending an exhibition at the National Portrait Gallery at around this time, Grace Chisholm Young was amused that she and her lady companion were stopped and searched, presumably for hammers or other items that could inflict damage. LUSA, Young Papers, D140/7/1 (Grace letter, n.d., 1912).

49   These banners are held as part of the *Women's Library Suffrage Banners Collection* in London.

50   Fawcett was carried three times around this bonfire on the shoulders of her fellow students, amid shouts of triumph and celebration.

51   RSL, Letter 948 (Huggins to Larmor, November 2 1906).

52   Shapin and Schaffer, pp. 22–79.

53   RSL, Referee Report 148 (George Carey Foster, 1901). Carey-Foster (1835–1919) was Principal of University College, London, a position that he had assumed in 1898, previous to which he had been Professor of Experimental Physics.

54   Sir Horace Lamb (1849–1934) Professor of Mathematics at Manchester University, author of texts on acoustics and fluid dynamics; John Joly (1857–1933) Professor of Geology and Mineralogy, Trinity College, Dublin.

55 RSL, Referee Report 142 (Horace Lamb, 1904) and 143 (John Joly, 1904).
56 Independent experiments in the 1950s corroborated Hertha's findings, although they made no reference to her work. See Tattersall and McMurran, p. 103.
57 IEE, Trotter, p. 590.
58 Marie-Clare Balaam, 'Representations of menopause and menopausal women in turn-of-the-century British medical journals', *Women's History Notebooks*, 7 (1) (2000), 10–14 (pp. 11–12).
59 Sir Almroth Wright, 'Militant hysteria'.
60 RSL, Letters 91 (April 12 1910); 331 (February 11 1911); 347 (February 13 1911); 394 (March 2 1911); 481 (April 5 1911). Letter 331 is from The Royal Society's Asst. Sec., Prof. Harrison, to Prof. J.H. Poynting (Hertha's 'communicator'), asking 'whether you really think it expedient to place it (referee report) in the hands of Mrs Ayrton, and whether it would not be likely to lead only to a controversy instead of to her acceptance of the Committee's decision ... that the descriptive part only of the paper ... be recommended for publication, and nothing was said about furnishing her with the grounds on which their decision was arrived at'. Other disputes amongst male protagonists had caused controversy and the Society did not wish to see this repeated. Poynting (1852–1914) was Professor of Physics at Mason College, Birmingham.
61 IEE, Trotter, p. 585.
62 In a lecture to Girton College in 1920, Hertha described her invention: 'The vortices are produced by flapping one plane on another many times in rapid succession, and it was found that the most efficient fan was constructed by hinging together two flat services – one flexible, of some such material as canvas, and the other rigid, perhaps of wood. The wooden part had to be held firm by a handle or other device, while the rest of the apparatus was flapped on to it by working up and down a stick attached to the canvas'. Ilse F. Stearn, 'A war-time invention: Lecture by Mrs Ayrton, July 27th', *Girton Review*, Jubilee supplement (1920), 42–43.
63 Kings College, London, The Liddle Hart Centre for Military Archives, The Foulkes Archive (Records of Central Gas Laboratory).
64 IEE, Trotter, p. 587.
65 Hertha Ayrton, 'Anti-gas fans', p. 8.
66 Major General Charles Howard Foulkes, '*Gas!' The Story of the Special Brigade*, London/Blackwell, 1934, p. 102.
67 Balaam, p. 12. This was an extreme view.
68 Lisa Tickner, 'Suffrage campaigns: The political imagery of the British women's suffrage movement', in *The Edwardian Era*, ed. by Jane Beckett and Deborah Cherry (Oxford: Phaidon, 1987), pp. 100–117.
69 Huggins, pp. 38–60 (p. 43).
70 Rose, p. 120.
71 The position of women in the BAAS remained contentious however. In the early 1920s there was argument over the election of botanist Agnes Arber as a section president with some fellows opining 'that there has been too much indirect influence of women in the botanical world through matrimony for some years, and its does not make for robustness' and 'the Council has been misled by a woman whose myopic vision does not extend beyond the

ringfences of Newnham'. Glasgow University, Archive Services, Papers of Frederick Orpen Bower, GB 0248 DC 002/14/18 (letter from Sir Isaac Bayley Balfour to Frederick Orpen Bower, January 20 1921) and GB 0248 DC 002/14/24 (letter from Frederick Orpen Bower to William Abbot Herdman, January 24 1921).

72 Evelleen Richards, 'Redrawing the boundaries: Darwinian science and Victorian women intellectuals', in *Victorian Science in Context*, ed. by Bernard Lightman (Chicago: University of Chicago Press, 1997), pp. 119–142 (p. 126).

73 Marilyn Bailey Ogilvie, 'Obligatory amateurs: Annie Maunder (1868–1947) and British women astronomers at the dawn of professional astronomy', *British Journal for the History of Science*, 33 (1) (2000), 67–84 (p. 77).

74 'The uses of a line-divider' (1885) and 'Experiments on the production of sand ripples' (1907).

75 For the founding of the Physical Society of London see Gooday, 'Teaching Telegraphy', pp. 80–85.

76 IEE, Hertha Ayrton election sheet. Hertha was nominated by Joseph Swan and Sylvanus P. Thompson, seconded by John Perry and others.

77 IEE, *Gertrude Entwisle* <http://www.iee.org/TheIEE/Research/Archives/Histories&Biographies/Entwisle.cfm> Entwisle (1892–1961) studied physics at Manchester with Rutherford and then transferred to engineering. In 1915 she joined Vickers Electrical Company; she was a founder member in 1919 of the Women's Engineering Society. Her obituary by 'I.H.H.', in *The Times* described her as 'in 1954 the first woman in Britain to retire from a complete career in industry as a professional engineer': I.H.H., 'Miss Gertrude Entwisle', *The Times*, November 27 1961, Obituaries, p. 18.

78 'A lady entomologist', *Nature*, 70 (October 13 1904), Book Reviews, p. 219.

79 For example: 'Two Fellows of the Royal Geographical Society', 'The Royal Geographical Society and women', *The Times*, August 6 1892, Letters to the Editor, p. 7 and William Hicks, 'The Royal Geographic Society and women', June 10 1893, Letters to the Editor, p. 12. The Council of the Society at one stage issued a circular disavowing the actions of the Secretary; one battle in a long-running dispute.

80 Mason, 'Hertha Ayrton and Royal Society', reaches this conclusion too.

81 RSL, Letter 10010 (Hertha Ayrton to Archibald Geikie, November 26 1910).

82 'From a correspondent' (Marie Stopes), 'Women and science: Complaint against the Royal Society', *The Times*, June 16 1914, News, p. 5. A short response was made by Muriel Robertson (1883–1973) of the Lister Institute of Preventative Medicine, who in a letter defended the Royal Society's equal treatment of women in all respects apart from the Fellowship: *The Times*, June 20 1914, Letters, p. 11.

83 Information held at News International Archive and Record Office, London E98 1ES.

84 There were precedents for exclusively male scientific institutions to award women prizes. Sophie Germain had won the grand prize of the Parisian Académie in 1816 for her work on elasticity; mathematician Sophia Kovalevskaia won the Bordin Prize of the French Academy of Sciences in 1888, although they would not admit her as a fellow.

85 RSL, Medal Claims, 1905 (John Perry for Hertha Ayrton, p. 267).

86 Ibid., 1906 (John Perry for Hertha Ayrton, pp. 290–291). Presumably Perry was including student and associate members to arrive at this number for IEE membership.
87 The conflicting requirements of politics and diplomacy may have been the reason why Larmor's candidate was seconded by John Perry, Hertha's prime nominator.
88 Zangwill, p. 21.
89 Sharp, *Hertha Ayrton*, p. 182.
90 Harvey, p. xi.
91 Mozans, pp. 386–387.

Conclusion

1 Women Into Science, Engineering and Mathematics (WISE) is a UK-based organisation which works with various groups to attract more girls and women to careers in science, engineering, construction and mathematics.
2 Wilson, 'Experimentalists among the mathematicians', p. 133.
3 C.P. Snow, *The Two Cultures, with an Introduction by Stefan Collini* (Cambridge: Cambridge, University Press, 1993). This book, originally published in 1964, originated from a 1959 lecture.

Bibliography

Archival Collections

Cambridge, Girton College Library and Archive: Bodichon Papers, manuscript by Joan Mason, 'Matilda Chaplin Ayrton (1846–1883), William Edward Ayrton (1847–1908) and Hertha Ayrton (1854–1923)'

Exeter University, Library (Special Collections): Correspondence Hertha Ayrton and Norman Lockyer

Glasgow University, Archive Services, Papers of Frederick Orpen Bower (correspondence GB 0248 DC 002/14/18 and GB 0248 DC 002/14/24)

Institution of Electrical Engineers, Library and Archives: Election sheet (Hertha Ayrton), memoirs of A.P. Trotter, correspondence Hertha Ayrton and Sylvanus P. Thompson

Liverpool Patent Office: Patent certificates for Phoebe Sarah Marks and Hertha Ayrton

Liverpool University, Special Collections and Archives: Papers of Dr. Rosalind Cecilia Hildegard Tanner, including papers of her parents W.H. Young and Grace Chisholm Young (GB 141 D599); Papers of Professor W.H. Young and of his wife Grace Chisholm Young (D140)

London Guildhall Library, Department of Manuscripts: Records of Finsbury Technical College, Records of the Central College of Technology, South Kensington

London, Kings College, The Liddle Hart Centre for Military Archives: The Foulkes Archive

London, News International, Archive and Record Office: Marie Stopes article

Manchester High School for Girls Archive: Burstall Testimonial (L 1886 1)

Paris, Bibliothèque nationale: Curie Papers (NAF 18443, fol. 301–324)

Royal Holloway, University of London, Archive and Records: Bedford College Magazine; correspondence Ethel Maude Rowell

Royal Society of London, Library, Archives and Manuscripts: Minutes of Council, Royal Society medal claims, Larmor/Huggins correspondence, conversazione programmes, Royal Society yearbooks, referee reports, letter books, Hertha Ayrton certificate of candidature, scrapbooks, press cuttings and ephemera

Smith College, Massachusetts, U.S.A.: Sophia Smith Collection, Mary Frances Newson Winston Papers

University College London, Library Manuscripts Room: Papers and Correspondence of Sir Francis Galton, 1822–1911

Women's Library: Autograph Letter Collection, correspondence Hertha Ayrton to Mrs. Gorthorn

Published Sources

Abbott, Edwin A., *Flatland: A Romance of many Dimensions*, 6th edn (Oxford: Blackwell, 1950)

Abir-am, Pnina G. and Outram, Dorinda, eds, *Uneasy Careers and Intimate Lives: Women in Science, 1789–1979* (New Brunswick, NJ: Rutgers University Press, 1987)

Adams, Pauline, *Somerville for Women: An Oxford College 1897–1993* (Oxford: Oxford University Press, 1996)

Albers, Donald J., Alexanderson, G.L. and Reid, Constance, *International Mathematical Congresses: An Illustrated History 1893–1986* (Berlin: Springer-Verlag, 1987)

Albery, David and Schwartz, Joseph, *Partial Progress: The Politics of Science and Technology* (London: Pluto, 1982)

Albisetti, James C., *Schooling German Girls and Women: Secondary and Higher Education in the Nineteenth Century* (Princeton: Princeton University Press, 1988)

Ameriks, Karl, ed., *The Cambridge Companion to German Idealism* (Cambridge: Cambridge University Press, 2000)

Andrews, A.E., 'A Statistical Study of Women Mathematicians in the Six Editions of "American Men of Science"', *Journal of Educational Sociology*, 17 (9) (1944), 543–550

Andrade, E.N. Da C., *A Brief History of the Royal Society* (London: The Royal Society, 1960)

Ardener, Shirley, ed., *Women and Space: Ground Rules and Social Maps,* 2nd edn (Oxford: Berg, 1993)

Argles, Michael, *South Kensington to Robbins: An Account of English Technical and Scientific Education since 1851* (London: Longman, 1964)

Armstrong, Henry E., 'Mrs Hertha Ayrton', *Nature,* 112 (December 1 1923), 800–801

Armstrong, Henry E., 'Technical education in Ireland', *The Times,* June 11 1901, Letters to Editor, p. 12

Ashworth, William J., 'Memory, efficiency and symbolic analysis: Charles Babbage, John Herschel and the industrial mind', *ISIS,* 87 (4) (1996), 629–653

Atkinson, Dwight, *Scientific Discourse in Sociohistorical Context: The Philosophical Transactions of the Royal Society of London, 1675–1975* (London: L. Erlbaum, 1999)

Atkinson, Paul, 'Fitness, feminism and schooling', in *The Nineteenth Century Woman: Her Cultural and Physical World,* ed. by Sara Delamont and Lorna Duffin, (London: Croom Helm, 1978), pp. 92–133

Ayrton, Hertha, 'Anti-gas fans: Their utility in France', *The Times,* May 10 1920, Letters to Editor, p. 8

Ayrton, Hertha, *The Electric Arc* (London: The Electrician Printing and Publishing Company, 1903)

Ayrton, W.E., *Practical Electricity: A Laboratory and Lecture Course,* 2nd edn (London: Cassell, 1900)

Balaam, Marie-Clare, 'Representations of menopause and menopausal women in turn-of-the-century British medical journals', *Women's History Notebooks,* 7 (1) (2000), 10–14

Baldick, Robert, *The Duel: A History of Duelling* (London: Chapman and Hall, 1965)

Barrow-Green, June, '"A corrective to the spirit of too exclusively pure mathematics": Robert Smith (1689–1768) and his prizes at Cambridge University', *Annals of Science,* 56 (3) (1999), 271–316

Barton, Ruth, '"Just before *Nature*": The purposes of science and the purposes of popularisation in some English popular science journals of the 1860s', *Annals of Science,* 55 (1998), 1–55

Battersby, Christine, *Gender and Genius: Towards a Feminist Aesthetics* (London: Women's Press, 1994)

Becker, B.J., 'Dispelling the myth of the able assistant: Margaret and William Huggins at work in the Tulse Hill Observatory', in *Creative Couples in the Sciences*, ed. by H.M. Pycior, N.G. Slack and P.G. Abir-am (New Brunswick, NJ: Rutgers University Press, 1996), pp. 98–111

Beckman, Linda Hunt, '"Leaving the tribal duckpond": Amy Levy, Jewish self-hatred and Jewish identity', *Victorian Literature and Culture*, 27 (1) (1999), 185–201

Bedford College for Women Calendar, 1899, 1910–11

Beetham, Margaret, *A Magazine of her Own? Domesticity and Desire in the Women's Magazine* (London: Routledge, 1996)

Benjamin, M., '"Elbow room": Women Writers on Science, 1790–1840', in *Science and Sensibility: Gender and Scientific Enquiry 1780–1945*, ed. by M. Benjamin (Oxford: Basil Blackwell, 1991), pp. 27–59

Bloor, David, 'Formal and informal thought', in *Science in Context: Readings in the Sociology of Science*, ed. by Barry Barnes and David Edge (Milton Keynes: Open University Press, 1982), pp. 117–124

Boyer, Carl B., *A History of Mathematics* (New York: Wiley, 1968)

Bradbrook, M.C., *'That infidel place': A Short History of Girton College, 1869–1969* (London: Chatto and Windus, 1969)

Bremner, C.S., *Education of Girls and Women in Great Britain* (London: Sonnenschein, 1897)

Brock, W.H., 'Building England's first technical college: The laboratories of Finsbury Technical College, 1878–1926', in *The Development of the Laboratory: Essays on the Place of Experiment in Industrial Civilization*, ed. by Frank A.J.L. James, (Basingstoke: Macmillan, 1989), pp. 154–170

Broks, Peter, *Media Science before the Great War* (Basingstoke: Macmillan, 1996)

Brück, Mary T., 'Alice Everett and Annie Russell Maunder, torch bearing women astronomers', *Irish Astronomical Journal*, 21 (3/4) (1994), 281–291

Burstall, S.A., 'The place of mathematics in girls' education', *The Mathematical Gazette*, 6 (96) (1912), 203–213

Burstall, Sara A., *Retrospect and Prospect: Sixty Years of Women's Education* (London: Longmans, Green and Co., 1933)

Burstyn, Joan N., *Victorian Education and the Ideal of Womanhood* (London: Croom Helm, 1980)

Caine, Barbara, *Destined to be Wives: The Sisters of Beatrice Webb* (Oxford: Clarendon Press, 1986)

Cambridge Philosophical Society, *Proceedings of the Cambridge Philosophical Society*, 4–17 (1880–1912/14)

Cardwell, D.S.L., *The Organisation of Science in England* (London: Heinemann, 1972)

Cartwright, Mary L., 'Grace Chisholm Young', *Girton Review*, Spring (1947), 17–19

Casteras, Susan P., 'The cult of the male genius in Victorian painting', in *Rewriting the Victorians: Theory, History and the Politics of Gender*, ed. by Linda M. Shires, (New York: Routledge, 1992), pp. 116–146

Chisholm, Grace, 'Extracts from a letter to the mathematical club', *Girton Review*, 37 (1894), 1–4

Clark, Beverly Lyon, 'Of snarks and games ... and publishing', *Children's Literature Association Quarterly*, 16 (2) (1991), 91–92

Clegg, Brian, *Infinity: The Quest to Think the Unthinkable* (London: Robinson, 2003)

Cockburn, Cynthia, 'Technology, production and power', in *Inventing Women: Science, Technology and Gender*, ed. by Gill Kirkup and Laurie Smith Keller, (Milton Keynes: Open University Press, 1992), pp. 196–211

Cohen, ed., *Talk on the Wilde Side: Toward a Genealogy of Discourse on Male Sexualities* (New York: Routledge, 1993)

Conrad, Joseph, *Heart of Darkness* (London: Penguin, 1994)

Corelli, Marie, *Woman or – Suffragette? A Question of National Choice* (London: C. Arthur Pearson, 1907)

Cornford, F.M. and Johnson, Gordon, *University Politics, F.M. Cornford's Cambridge and his Advice to the Young Academic Politician, containing the complete text of Cornford's 'Microcosmographia Academica'* (Cambridge: Cambridge University Press, 1994)

Creese, Mary R.S. 'Elles, Gertrude Lilian (1872–1960)', *Oxford Dictionary of National Biography* (Oxford: Oxford University Press, 2004)

Creese, Mary R.S., 'Saunders, Edith Rebecca (1865–1945)', *Oxford Dictionary of National Biography* (Oxford: Oxford University Press, 2004)

Creese, Mary R.S., *Ladies in the Laboratory? American and British Women in Science, 1800–1900: A Survey of their Contributions to Research* (London: Scarecrow Press, 1998)

David, Deirdre, '"Art's a service": Social wound, sexual politics and Aurora Leigh', in *Victorian Women Poets: Emily Brontë, Elizabeth Barrett Browning, Christina Rossetti*, ed. by Joseph Bristow (Basingstoke: Macmillan, 1995), pp. 108–131

David, Deirdre, *Intellectual Women and Victorian Patriarchy* (Basingstoke: Macmillan, 1987)

Davies, Emily, 'Special systems of education for women (1868)', in *The Education Papers: Women's Quest for Equality in Britain, 1850–1912*, ed. by Dale Spender (New York: Routledge and Kegan Paul, 1986), pp. 99–122

Davis, A.E.L., 'Hardcastle, Frances (1866–1941)', *Oxford Dictionary of National Biography* (Oxford: Oxford University Press, 2004)

Delamont, Sara, *Knowledgeable Women: Structuralism and the Reproduction of Elites* (London: Routledge, 1989)

Delap, Lucy, 'The superwoman: Theories of gender and genius in Edwardian Britain', *Historical Journal*, 47 (1) (2004), 101–126

Delap, Lucy, '"Philosophical vacuity and political ineptitude": *The Freewoman's* critique of the suffrage movement', *Women's History Review*, 11 (4) (2002), 613–630

De Morgan, Augustus, 'Presidential address', *Proceedings of the London Mathematical Society*, 1 (1865–1866), 1–4

Diethe, Carol, *Nietzsche's Women: Beyond the Whip* (Berlin: Walter de Gruyter, 1996)

Dyhouse, Carol, *No Distinction of Sex? Women in British Universities, 1870–1939* (London: UCL Press, 1995)

Dyhouse, Carol, *Girls Growing up in Late Victorian and Edwardian England* (London: Routledge and Kegan Paul, 1981)

Dyhouse, Carol, 'Good wives and little mothers: Social anxieties and the schoolgirl's curriculum, 1890–1920', *Oxford Review of Education*, 3 (1) (1977), 21–35

Eagleton, Terry, 'The flight to the real', in *Cultural Politics at the Fin de Siècle*, ed. by Sally Ledger and Scott McCracken (Cambridge: Cambridge University Press, 1995), pp. 11–21

Elliott, E.B., 'Some secondary needs and opportunities of English mathematicians', *Proceedings of the London Mathematical Society*, 30 (1898–1899), 5–23

Erskine, Fiona, 'The *Origin of Species* and the science°of female inferiority', in *Darwin's 'The Origin of Species': New Interdisciplinary Essays*, ed. by David Amigoni and Jeff Wallace (Manchester: Manchester University Press, 1995), pp. 95–121

Eschbach, Elizabeth Seymour, *The Higher Education of Women in England and America, 1865–1920* (New York: Garland, 1993)

Eyre, J. Vargas, *Henry Edward Armstrong, 1848–1937* (London: Butterworth Scientific, 1958)

Falconer, Isobel, 'J.J. Thomson and "Cavendish Physics"', in *The Development of the Laboratory: Essays on the Place of Experiment in Industrial Civilization*, ed. by Frank A.J.L. James (Basingstoke: Macmillan, 1989), pp. 104–117

Fenster, Della Dumbaugh and Parshall, Karen Hunger, 'Women in the American Mathematical Research Community: 1891–1906', in *The History of Modern Mathematics*, vol. III, *Images, Ideas and Communities*, ed. by E. Knobloch and D. Rowe (San Diego: Academic Press, 1994)

Fletcher, Sheila, 'The making and breaking of a female tradition: Women's physical education in Britain, 1880–1980', *British Journal of Sports History*, 2 (1985), 29–39

Forfar, D.O., 'What became of the Senior Wranglers?', *Mathematical Spectrum*, 29 (1) (1996)

Forsyth, A.R., 'Old tripos days at Cambridge', *Mathematical Gazette*, 29 (1935), 162–179

Forsyth, A.R., 'Obituary of James Whitehead Lee Glaisher, 1848–1928', *Proceedings of the Royal Society of London, Series A*, 26 (1930), 1–11

Foulkes, C.H., *'Gas!' The Story of the Special Brigade* (Edinburgh: Blackwood, 1934)

Fox Keller, Evelyn, *Reflections on Gender and Science* (New Haven: Yale University Press, 1985)

Franks, Rev. Dr. R.S., 'Mr. Robert Webb', *The Times*, August 5 1890, Obituaries, p. 14

Frevert, Ute, *Women in German History: From Bourgeois Emancipation to Sexual Liberation*, trans. by Stuart McKinnon-Evans, Terry Bond and Barbara Norden (Oxford: Berg, 1986)

Frevert, Ute, 'The civilizing tendency of hygiene: Working class women under medical control in Imperial Germany', in *German Women in the Nineteenth Century: A Social History*, ed. by John C. Fout (New York: Holmes and Meier, 1984), pp. 320–344

Furinghetti, Fulvia, 'The emergence of women on the international stage of mathematics education', *ZDM Mathematics Education*, 40 (529) (2008), 529–543

Gardner, Alice, *A Short History of Newnham College* (Cambridge: Bowes, 1921)

Garner, Lesley, *Stepping Stones to Women's Liberty* (London: Heinemann, 1984)

Garrard, John A., *The English and Immigration, 1880–1910* (Oxford: Oxford University Press, 1971)

Gay, Hannah, 'Invisible resource: William Crookes and his circle of support, 1871–81', *British Journal for the History of Science*, 29 (1996), 311–336

Geikie, Sir Archibald, *Annals of the Royal Society Club: The Record of a London Dining-club in the Eighteenth and Nineteenth Centuries* (London: Macmillan, 1917)

Gilmour, Robin, *The Victorian Period: The Intellectual and Cultural Context of English Literature, 1830–1890* (Harlow: Longmans, 1993)

Girton College, *Girton College Register*, 1869–1946, ed. by K.T. Butler and H.I. McMorran (Cambridge: privately printed, 1948)

Girton Review, 'Margaret Theodora Meyer', May (1924), 6–7

Girton Review, 'The mathematical tripos', December (1882), 3–4

Glaisher, J.W.L., 'Presidential address, section A: Mathematical and Physical Science', in *Report of the 60th Meeting of the British Association for the Advancement of Science, Leeds, 1890* (London: British Association, 1891), pp. 719–727

Gooday, Graeme J.N., 'Ayrton, William Edward (1847–1908)', *Oxford Dictionary of National Biography* (Oxford: Oxford University Press, 2004) [http://www.oxforddnb.com/view/article/30509, accessed 29 January 2005]

Gooday, Graeme J.N., *The Morals of Measurement: Accuracy, Irony, and Trust in Late Victorian Electrical Practice* (Cambridge: Cambridge University Press, 2004)

Gooday, Graeme J.N., 'The premisses of premises: Spatial issues in the historical construction of laboratory credibility', in *Making Space for Science: Territorial Themes in the Shaping of Knowledge*, ed. by Crosbie Smith and John Agar (Basingstoke: Macmillan, 1998), pp. 216–245

Gooday, Graeme J.N., 'Faraday reinvented: Moral imagery and institutional icons in Victorian electrical engineering', *History of Technology*, 15 (1993), 190–205

Gooday, Graeme J.N., 'Teaching telegraphy and electrotechnics in the physics laboratory: William Ayrton and the creation of an academic space for electrical engineering in Britain, 1873–1884', *History of Technology Journal*, 13 (1991), 73–111

Gooday, Graeme, 'Precision measurement and the genesis of physics teaching laboratories in Victorian Britain', *British Journal for the History of Science*, 23 (1990), 25–51

Gooding, David, Pinch, Trevor and Schaffer, Simon, eds, *The Uses of Experiment: Studies in the Natural Sciences* (Cambridge: Cambridge University Press, 1989)

Gould, Paula, 'Women and the culture of university physics in late nineteenth-century Cambridge', *British Journal for the History of Science*, 30 (2) (1997), 127–149

Grattan-Guinness, I., 'A mathematical union: William Henry Young and Grace Chisholm', *Annals of Science*, 29 (2) (1972), 105–186

Grattan-Guinness, I., 'University mathematics at the turn-of-the-century: Unpublished recollections of W.H. Young', *Annals of Science*, 28 (4) (1972), 367–384

Gray, Jeremy J., 'Anxiety and abstraction in nineteenth-century mathematics', *Science in Context*, 17 (1/2) (2004), 23–47

Green, Judy and LaDuke, Jeanne, 'Women in American mathematics: A century of contributions', in *A Century of Mathematics in America, Part II*, ed. by Peter Duren with the assistance of Richard A. Askey and Uta C. Merzbach (American Mathematical Society, 1989), pp. 379–398

Green, Judy and LaDuke, Jeanne, 'Women in the American mathematical community: The pre-1940 Ph.D.'s', *The Mathematical Intelligencer*, 9 (1) (1987), 11–23

Green, Martin Burgess, *Dreams of Adventure: Deeds of Empire* (London: Routledge and Kegan Paul, 1980)

Grier, David Alan, *When Computers Were Human* (Princeton: Princeton University Press, 2005)

Grinstein, Louise S., Rose, Rose K. and Rafailovich, Miriam H., eds, *Women in Chemistry and Physics: A Biobibliographic Sourcebook* (Connecticut and London: Greenwood Press, 1993)

Haggard, Rider H., *King Soloman's Mines* (Hertfordshire: Wordsworth Editions, 1998)

Halberstam, Judith, 'Technologies of monstrosity: Bram Stoker's Dracula', in *Cultural Politics at the Fin de Siècle*, ed. by Sally Ledger and Scott McCracken (Cambridge: Cambridge University Press, 1995), pp. 248–266

Hall, Rupert A., *The Cambridge Philosophical Society: A History, 1819–1969* (Cambridge: Cambridge Philosophical Society, 1969)

Hamerton, Philip G., *The Intellectual Life* (London: Macmillan, 1911)

Hamilton, Mary Agnes, *Newnham: An Informal Biography* (London: Faber and Faber, 1936)

Hardy, G.H., *A Mathematician's Apology, with a forward by C.P. Snow*, 2nd edn (Cambridge: Cambridge University Press, 2001)

Hardy, G.H., 'W.H. Young', *Journal of the London Mathematical Society*, 17 (1942), 218–237

Hardy, G.H., 'William Henry Young', *Nature*, August 22 1942, Obituaries, pp. 27–28

Hargraves, Jennifer A., '"Playing like gentlemen while behaving like ladies": Contradictory features of the formative years of women's sport', *British Journal of Sports History*, 2 (1985), 40–52

Harman, Peter M., *The Scientific Letters and Papers of James Clerk Maxwell*, 3 vols (Cambridge: Cambridge University Press, 1990–2003)

Harte, Negley, *The University of London 1836–1986: An Illustrated History* (London: The Athlone Press, 1986)

Harte, Negley Boyd and North, John, *The World of University College, London, 1828–1978, with Introduction by Lord Annon* (London: University College, London, 1979)

Harvey, Joy, *'Almost a Man of Genius': Clémence Royer, Feminism and Nineteenth-century Science* (New Brunswick, NJ: Rutgers University Press, 1997)

Heard, John, 'Of apes and algebraists', *Viewpoint: Newsletter of the British Society for the History of Science*, 83 (June 2007), 1–3

Heathorn, Stephen, '"The highest type of Englishman": Gender, war and the Alfred the Great millenary commemoration of 1901', *Canadian Journal of History*, 37 (2002), 459–484

Hicks, William, 'The royal geographical society and women', *The Times*, June 10, 1893, Letters to the Editor, p. 12

Hobson, E.W. and Love, A.E.H., eds, *Proceedings of the Fifth International Congress of Mathematicians … 1912* (2 vols) (Cambridge: Cambridge University Press, 1913)

Hoffmann, Stefan-Ludwig, 'Civility, male friendship and Masonic sociability in nineteenth-century Germany', *Gender and History*, 13 (2) (2001), 224–248

Howarth, Janet, '"In Oxford but … not of Oxford": The Women's colleges', in *The History of the University of Oxford* (Vol. VII, Nineteenth-Century Oxford, Part 2) ed. by M.G. Brocks and M.C. Curthoys (Oxford: Clarendon Press, 2000), pp. 237–307

Howarth, Janet, '"Oxford for Arts": The natural sciences, 1880–1914', in *The History of the University of Oxford* (Vol. VII, Nineteenth-Century Oxford, Part 2) ed. by M.G. Brocks and M.C. Curthoys (Oxford: Clarendon Press, 2000), pp. 457–497

Howarth, Janet, 'Science education in late-Victorian Oxford: A curious case of failure?', *The English Historical Review*, 102 (403) (1987), 334–371

Howarth, O.U.R., *The British Association for the Advancement of Science: A Retrospect, 1831–1931* (London: British Association for the Advancement of Science, 1931)

Howson, Geoffrey, *A History of Mathematical Education in England* (Cambridge: Cambridge University Press, 1982)

Hudson, Hilda P., 'Mathematics and eternity', *The Mathematical Gazette*, XII (174) 1925, 265–270

Huggins, Sir William, *The Royal Society, or, Science in the State and in the Schools* (London: Methuen, 1906)

Hull, Andrew, '"War of words": The public science of the British scientific community and the origins of the Department of Scientific and Industrial Research, 1914–16', *British Journal for the History of Science*, 32 (4) (1999), 461–481

Humphrey, A.T., 'Lord Rayleigh: The last of the great Victorian polymaths', *GEC Review*, 7 (3) (1992), 167–180

I.H.H., 'Miss Gertrude Entwisle', *The Times*, November 27 1961, Obituaries, p. 18

Iaglom, Isaak Moiseevich, *Felix Klein and Sophus Lie: Evolution of the Idea of Symmetry in the Nineteenth Century*, trans. by Sergei Sassinsky, Hardy Grant and Abe Shenitzer (Boston: Birkhäuser, 1988)

James, Frank A.J.L., ed., *The Development of the Laboratory: Essays on the Place of Experiment in Industrial Civilization* (Basingstoke: Macmillan, 1989)

James, Ioan, *Remarkable Mathematicians: From Euler to von Neumann* (Cambridge: Cambridge University Press, 2002)

Jardine, Nicholas, *The Scenes of Inquiry: On the Reality of Questions in the Sciences* (Oxford: Clarendon Press, 1991)

John, Angela V., '"Behind the locked door": Evelyn Sharp, suffragette and rebel journalist', *Women's History Review*, 12 (1) (2003), 5–13

Jones, Claire, 'Grace Chisholm Young: Gender and mathematics around 1900', *Women's History Review*, 9 (4) (2000), 675–692

Jones, Greta, 'Bell, Julia (1879–1979), geneticist', *Oxford Dictionary of National Biography* (Oxford: Oxford University Press, 2004)

Jones, Greta, *Social Hygiene in Twentieth-century Britain* (London: Croom Helm, 1986)

Jordanova, Ludmilla, *Sexual Visions: Images of Gender in Science and Medicine between the Eighteenth and Twentieth Centuries* (Wisconsin: University of Wisconsin Press, 1993)

Jungwirth, Helga, 'Reflections on the foundations of research on women and mathematics', in *Math Worlds: Philosophical and Social Studies of Mathematics and Mathematics Education*, ed. by Sal Restivo, Jean Paul Bendegem and Roland Fisher (Albany: State University of New York Press, 1993), pp. 134–149

Kanitkar, Helen, '"Real true boys": Moulding the cadets of Imperialism', in *Dislocating Masculinity: Comparative Ethnographies*, ed. by Andrea Cornwell and Nancy Lindisfarne (London: Routledge, 1994), pp. 184–196

Kennedy, Ellen, 'Nietzsche: Women as Untermensch', in *Women in Western Political Philosophy: Kant to Nietzsche*, ed. by Ellen Kennedy and Susan Mendus (Brighton: Wheatsheaf, 1987), pp. 179–201

Kenschaft, Patricia C., 'Charlotte Angas Scott, 1858–1931', *College Mathematics Journal*, 18 (2) (1987), 98–110

Kenschaft, Patricia C., 'Charlotte Angas Scott, 1858–1931', in *Women of Mathematics*, ed. by Louise S. Grinstein and Paul J. Campbell (Westport, Conn: Greenwood, 1987), pp. 193–203

Kenschaft, Patricia C., 'Women in mathematics around 1900', *Signs*, 7 (4) (1982), 906–909

Kesiner, J.A., *Sherlock's Men: Masculinity, Conan Doyle and Cultural History* (Aldershot: Ashgate, 1997)

Kessel, Martina, 'The whole man: The longing for a masculine world in nineteenth-century Germany', *Gender and History*, 15 (1) (2003), 1–31

Kidwell, Peggy Aldrich, 'Women astronomers in Britain, 1780–1930', *ISIS*, 75 (3) (1984), 534–546

Kilmister, Clive, 'Genius in mathematics', in Murray, Penelope A., ed., *Genius: The History of an Idea* (Oxford: Blackwell, 1989), pp. 181–195

Knott, Cargill G., ed., *Collected Scientific Papers of John Aitken, edited for the Royal Society of Edinburgh, with an Introductory Memoir by Cargill G. Knott* (Cambridge: Cambridge University Press, 1923)

Koblitz, Ann Hibner, *A Convergence of Lives. Sofia Kovalevskaia: Scientist, Writer, Revolutionary* (New Brunswick, N.J.: Rutgers University Press, 1993)

Lang, Jennifer, *City and Guilds of London Institute Centenary, 1878–1978* (London: City and Guilds of London Institute, 1978)

Ledger, Sally, *The New Woman: Fiction and Feminism at the Fin de Siècle* (Manchester: Manchester University Press, 1997)

Ledger, Sally and Roger, Luckhurst, *The Fin de Siècle: A Reader in Cultural History c. 1880–1900* (Oxford: Oxford University Press, 2000)

Linnaean Society, 'Mrs Hendrina (Rina) Victoria Scott', *Proceedings of the Linnaean Society* (1928–9), 146–7 (obituaries)

Lloyd, Genevieve, *The Man of Reason: 'Male' and 'Female' in Western Philosophy* (London: Methuen, 1984)

Lodge, Sir Oliver, *Advancing Science, Being Personal Reminiscences of the BA in the Nineteenth Century* (London: Benn, 1931)

London Mathematical Society, *Proceedings of the London Mathematical Society*, Series 1 (1865–1902) and Series 2 (1903/4–1914)

London University, *Calendar* (London: University of London, 1878–1914+)

Love, Rosaleen, 'Alice in eugenics-land: Feminism and eugenics in the careers of Alice Lee and Ethel Elderton', *Annals of Science*, 36 (2) (1979), 145–158

Mackenzie, Donald A., *Statistics in Britain 1865–1930: The Social Construction of Scientific Knowledge* (Edinburgh: Edinburgh University Press, 1981)

Mackenzie, Norman and Mackenzie, Jeanne, *The Life of H.G. Wells: The Time Traveller* (London: Hogarth Press, 1987)

Magnello, Eileen M., 'The non-correlation of biometrics and eugenics: Rival forms of laboratory work in Karl Pearson's career at University College London, part 1', *History of Science*, 37 (1) (1999), 79–106 and 'The non-correlation of biometrics ... part 2', *History of Science*, 37 (2) (1999), 123–150

Marks, Sarah, 'Abstracts from letters to Barbara Bodichon', *Girton Review* Michaelmas (1927), 8–11

Martin, Dan, 'Elizabeth A. McHarg', *Glasgow Mathematical Journal*, 42 (2000), 487–488

Mason, Joan, 'The women fellows' jubilee', *Notes and Records of the Royal Society of London*, 49 (1) (1995), 125–140

Mason, Joan, 'Hertha Ayrton and the admission of women to the Royal Society of London', *Notes and Records of the Royal Society of London*, 45 (2) (1991), 201–220

Mather, T., 'Mrs Hertha Ayrton', *Nature*, 112 (December 29, 1923), 939

Mayr, Otto, 'The science-technology relationship', in *Science in Context: Readings in the Sociology of Science*, ed. by Barry Barnes and David Edge (Milton Keynes: Open University Press, 1982), pp. 155–163

Mazón, Patricia M., *Gender and the Modern Research University: The Admission of Women to German Higher Education, 1865–1914* (California: Stanford University Press, 2003)

McMurran, Shawnee L. and Tattersall, James J., 'Mary Cartwright, 1900–1998', *Notices of the American Mathematical Society*, 46 (2) (1999), 214–220

McWilliams-Tullberg, Rita, 'Women and degrees at Cambridge University, 1862–1897', in *A Widening Sphere: Changing Roles of Victorian Women*, ed. by Martha Vicinus (Bloomington: Indiana University Press, 1977), pp. 117–145

McWilliams-Tullberg, Rita, *Women at Cambridge: A Men's University – Though of a Mixed Type* (London: Gollancz, 1975)

M.D.K., 'Hilda Phoebe Hudson, 1881–1965', *Newnham College Roll Letter* (1966), 53–54

Meadows, A.J., *Science and Controversy: A Biography of Sir Norman Lockyer* (London: Macmillan, 1972)

Mehrtens, Herbert, Bos, H.J.M. and Schneider, Ivo, eds, *Social History of Nineteenth Century Mathematics* (Boston: Birkhäuser, 1981)

Merchant, Carolyn, 'Isis' consciousness raised', in *History of Women in the Sciences: Readings from ISIS*, ed. by Sally Gregory Kohlstedt (Chicago: University of Chicago Press, 1999), pp. 11–22

Meschkowski, Herbert, *Ways of Thought of Great Mathematicians: An Approach to the History of Mathematics*, trans. by John Dyer-Bennet (San Francisco: Holden Day, 1964)

Messenger of Mathematics, 1–44 (1872–1914/15)

Morrison-Low, A.D., 'Women in the nineteenth-century scientific instrument trade', in *Science and Sensibility: Gender and Scientific Enquiry, 1780–1945*, ed. by M. Benjamin (Oxford: Basil Blackwell, 1991), pp. 89–117

Mozans, H.J., *Woman in Science, with an Introductory Chapter on Woman's Long Struggle for Things of the Mind*, 3rd edn (Notre Dame, Indiana: University of Notre Dame Press, 1991)

Nature, 'The laboratories of Bedford College', 61 (January 23, 1890), 279; 'On our bookshelves', 63 (November 8 1900), 28; 'A lady entomologist', 70 (October 13 1904), 219

Neeley, Kathryn A., *Mary Somerville: Science, Illumination and the Female Mind* (Cambridge: Cambridge University Press, 2001)

Newnham College, 'Hilda Phoebe Hudson, 1881–1965', *Newnham College Roll Letter* (1966), 53–54

Ogilvie, Marilyn Bailey, 'Obligatory amateurs: Annie Maunder (1868–1947) and British women astronomers at the dawn of professional astronomy', *British Journal for the History of Science*, 33 (1) (2000), 67–84

Ogilvie, Marilyn Bailey, *Women in Science, Antiquity through the Nineteenth Century: A Biographical Dictionary with Annotated Bibliography* (Cambridge, MA: MIT Press, 1991)

Oldfield, Sybil, 'Macmillan, (Jessie) Chrystal (1872–1937)', *Oxford Dictionary of National Biography*, Oxford University Press, 2004 <http://www.oxforddnb.com/view/article/38526> [accessed 27 March 2009]

Orton, James, *The Liberal Education of Women*, 2nd edn (New York: Barnes, 1896)

Osen, Lynn M., *Women in Mathematics* (Cambridge, MA: MIT Press, 1974)

Owens, Larry, 'Pure and sound government: Laboratories, playing fields and gymnasia in the nineteenth-century search for order', *ISIS*, 76 (1985), 182–194

Park, Jihang, 'The British suffrage activists of 1913: An analysis', *Past and Present*, 120 (1988), 147–162

Parshall, Karen Hunger and Rowe, David E., *The Emergence of the American Mathematical Research Community, 1876–1900: J.J. Sylvester, Felix Klein and E.H. Moore* (Providence RI: American Mathematical Society, 1991)

Patterson, Elizabeth Chambers, *Mary Somerville and the Cultivation of Science* (Boston, MA: Nijhoff, 1983)

Perrone, Fernanda, 'Women academics in England, 1870–1930', *History of Universities*, 12 (1) (1993), 339–367

Porter, Theodore M., *Trust in Numbers* (Princeton: Princeton University Press, 1995)

Pycior, Helena M., 'Pierre Curie and "His eminent collaborator Mme Curie": Complementary partners', in *Creative Couples in the Sciences*, ed. by Helena M. Pycior, Nancy G. Slack and Pnina G. Abir-Am (New Brunswick, NJ: Rutgers University Press, 1996), pp. 39–56

Pycior, Helena M., Slack, Nancy G. and Abir-Am, Pnina G., eds, *Creative Couples in the Sciences* (New Brunswick, NJ: Rutgers University Press, 1996)

Pyenson, Lewis, *Neohumanism and the Persistence of Pure Mathematics in Wilhelmian Germany* (Philadelphia: American Philosophical Society, 1983)

Quarterly Journal of Pure and Applied Mathematics, 1–45 (1857–1914)

Quinn, Susan, *Marie Curie: A Life* (Cambridge MA: Da Capo Press, 1995)

Raftery, Deirdre and Parkes, Susan M., eds, *Female Education in Ireland 1700–1900: Minerva or Madonna?* (Dublin: Irish Academic Press, 2007)

Rankin, R.A., 'The first hundred years (1883–1983)', *Proceedings of the Edinburgh Mathematical Society*, 26 (1983), 135–150

Rayleigh, Lord Robert John Strutt, *The Life of Sir J.J. Thomson* (Cambridge: Cambridge University Press, 1942)

Rayleigh, Lord Robert John Strutt, 'Some reminiscences of scientific workers of the past generation, and their surroundings', *Proceedings of the Physical Society*, 48 (2) (1936), 216–246

Rayleigh, Lord, 'Address of the president, Lord Rayleigh, O.M., D.C.L., at the anniversary meeting on November 30th, 1907', *Proceedings of the Royal Society of London, Series A*, 80 (1907–8), 239–251

Reader, W.L., *A History of the Institution of Electrical Engineers, 1871–1971* (London: Peregrinus, 1987)

Reid, Constance, *Hilbert* (London: Allen and Unwin, 1970)

Reid, Robert, *Marie Curie* (London: Collins, 1974)

Restivo, Sal, Van Bendegem, Jean Paul and Fischer, Roland, eds, *Math Worlds: Philosophical and Social Studies of Mathematics and Mathematics Education* (Albany: State University of New York Press, 1993)

Richards, Evelleen, 'Redrawing the boundaries: Darwinian science and Victorian women intellectuals', in *Victorian Science in Context*, ed. by Bernard Lightman (Chicago: University of Chicago Press, 1997), pp. 119–142

Richardson, Angelique, *Love and Eugenics in the late Nineteenth Century: Rational Reproduction and the New Woman* (Oxford: Oxford University Press, 2003)

Richmond, Marsha L., 'Women in the early history of genetics: William Bateson and the Newnham College mendelians, 1900–1910', *ISIS*, 92 (2001), 55–90

Richmond, Marsha L., '"A lab of one's own": The Balfour biological laboratory for women at Cambridge University, 1884–1914', in *History of Women in the Sciences: Readings from ISIS*, ed. by Sally Gregory Kohlstedt (Chicago: University of Chicago Press, 1999), pp. 235–268

Roberts, Samuel, 'Presidential address', *Proceedings of the London Mathematical Society*, 14 (1882–3), 6–12

Robertson, M., 'Science and women: A defence of the Royal Society, *The Times*, June 20, 1914, Letters, p. 11

Roper, Michael and Tosh, John, eds, *Manful Assertions: Masculinities in Britain since 1800* (London: Routledge, 1991)

Rose, Jonathan, *The Edwardian Temperament, 1895–1919* (Ohio: Ohio University Press, 1986)

Rosenheim, Shawn, '"The King of 'Secret Readers'": Edgar Poe, cryptography, and the origins of the detective story', *English Literary History*, 56 (2) (1989), 375–400

Rossiter, Margaret W., '"Women's work" in science', in *History of Women in the Sciences: Readings from ISIS*, ed. by Sally Gregory Kohlstedt (Chicago: University of Chicago Press, 1999), pp. 287–304

Rossiter, Margaret W., 'The Matthew/Matilda effect in science', *Social Studies of Science*, 23 (1993), 325–341

Rothblatt, Sheldon, *The Revolution of the Dons: Cambridge and Society in Victorian England* (Cambridge: Cambridge University Press, 1981)

Rothman, Patricia, 'Grace Chisholm Young and the division of the laurels', *Notes and Records of the Royal Society of London*, 50 (1) (1996), 89–100

Rowe, David E., 'Making mathematics in an oral culture: Göttingen in the era of Klein and Hilbert', *Science in Context*, 17 (1/2) (2004), 85–129

Rowe, David E., 'Klein, Hilbert and the Göttingen mathematical tradition', *Osiris*, sec. series, 5 (1989), 186–213

Rowe, David E., '"Jewish mathematics" at Göttingen in the era of Felix Klein', *ISIS*, 77 (3) (1986), 422–440

Rowold, Katharina, 'The many lives and deaths of Sofia Kovalevskaia: Approaches to women's role in scholarship and culture in Germany at the turn of the twentieth century', *Women's History Review*, 10 (4) 2001, 603–628

Royal Holloway College Calendar, 1897–1914

Royal Society of London, *The Celebration of the 250th Anniversary of the Royal Society, July 15–19, 1912* (London: Royal Society, 1913)

Royal Society of London, *Proceedings of, Series A, Containing Papers of a Mathematical and Physical Character*, 76–91 (1905–1915)

Royal Society of London, *Proceedings of, Series B, Containing Papers of a Biological Character*, 76–89 (1905–1915)

Royal Society of London, *Proceedings of, Containing Obituaries of Deceased Fellows Chiefly for the Period 1891–1904*, 75 (1905)

Royal Society of London, *Philosophical Transactions of, Series A, Containing Papers of a Mathematical or Physical Character*, 178–214 (1887–1914)

Royal Society of London, *Philosophical Transactions of, Series B, Containing Papers of a Biological Character*, 178–205 (1887–1914)

Royal Society of London, *Proceedings of*, 19–75 (1870–1905)

Royal Society of London, *Philosophical Transactions of*, 158–177 (1867–1886)

Russell, Bertrand, 'The study of mathematics', in *Mysticism and Logic, and Other Essays*, by Bertrand Russell (London: Penguin, 1953), pp. 58–73

Schaffer, Simon, 'Physics laboratories and the Victorian country house', in *Making Space for Science: Territorial Themes in the Shaping of Knowledge*, ed. by Crosbie Smith and John Agar (Basingstoke: Macmillan, 1998), pp. 149–180

Schaffer, Simon, 'Genius in romantic natural philosophy', in *Romanticism and the Sciences*, ed. by Andrew Cunningham and Nicholas Jardine (Cambridge: Cambridge University Press, 1990), pp. 82–98

Scheinberg, Cynthia, 'Re-mapping Anglo-Jewish literary history', *Victorian Literature and Culture*, 27 (1) (1999), 115–124

Schiebinger, Londa, *The Mind has no Sex? Women in the Origins of Modern Science* (Cambridge, MA: Harvard University Press, 1991)

Sciama, Lidia, 'Ambivalence and dedication: Academic wives in Cambridge University 1870–1970', in *The Incorporated Wife*, ed. by Hillary Callan and Shirley Ardener, (London: Croom Helm, 1984), pp. 50–66

Scott, Charlotte Angas, 'Paper read before the mathematical club at Girton College, May term, 1893', *Girton Review*, 36 (1984), 1–4

Sebeok, Thomas A. and Margolis, Harriet, 'Captain Nemo's porthole: Semiotics of windows in Sherlock Holmes', *Poetics Today*, 3 (1) (1982), 110–139

Semple, J.G., 'Hilda Phoebe Hudson', *Bulletin of the London Mathematical Society*, 1 (1969), 357–359

Shapin, Steven, *A Social History of Truth: Civility and Science in Seventeenth-century England* (Chicago: University of Chicago Press, 1994)

Shapin, Steven and Schaffer, Simon, *Leviathan and the Air-pump: Hobbes, Boyle and the Experimental Life* (Princeton: Princeton University Press, 1985)

Sharp, Evelyn, *Unfinished Adventure: Selected Reminiscences from an Englishwoman's Life* (London: Bodley Head, 1933)

Sharp, Evelyn, *Hertha Ayrton, 1854–1923: A Memoir* (London: Arnold, 1926)

Sharp, Evelyn, 'The late Mrs Hertha Ayrton', *The Times*, March 7 1925, Letters to Editor, p. 8

Shindler, Karolyn, *Discovering Dorothea: The Life of the Pioneering Fossil-Hunter Dorothea Bate* (London: HarperCollins, 2005)

Sidgwick, Ethel, *Mrs. Henry Sidgwick: A memoir by her Niece* (London: Sidgwick and Jackson, 1938)

Sidgwick, Mrs Henry, *Health Statistics of Women Students of Cambridge and Oxford and of their Sisters* (Cambridge: Cambridge University Press, 1890)

Sigurdsson, Skuli, 'Equivalence, pragmatic Platonism and the discovery of the calculus', in *The Invention of Physical Science: Intersections of Mathematics, Theology and Natural Philosophy since the Seventeenth Century*, ed. by Mary Jo Nye, Joan Richards, Roger H. Stuewer and Erwin N. Hiebert (Dordrecht: Kluwer Academic, 1992), pp. 97–116

Siklos, Stephen, *Philippa Fawcett and the Mathematical Tripos* (Cambridge: Newnham College, 1990)

Sime, Ruth Lewin, *Lise Meitner: A Life in Physics* (California: University of California Press, 1996)

Singer, Sandra L., *Adventures Abroad: North American Women at German-Speaking Universities, 1868–1915* (Westport, Conn.: Praeger 2003)

Smith, Crosbie W. and Wise, M. Norton, *Energy and Empire: A Biographical Study of Lord Kelvin* (Cambridge: Cambridge University Press, 1989)

Snow, C.P., *The Two Cultures, with an Introduction by Stefan Collini* (Cambridge: Cambridge University Press, 1993)

Solomon, Barbara Miller, *In the Company of Educated Women: A History of Women and Higher Education in America* (New Haven: Yale University Press, 1985)

Solomon, Maurice, 'Review of Evelyn Sharp's *Hertha Ayrton, 1854–1923: A Memoir*', *Central Gazette*, 23 (59) (1926), 70–72.

Spooner, Henry J., 'The education of engineers', *The Times*, Letters to Editor, July 19 1911, p. 24

Stanley, Autumn, *Mothers and Daughters of Invention: Notes for a Revised History of Technology* (New Brunswick: Rutgers University Press, 1995)

Staten, Henry, *Nietzsche's Voice* (Ithaca NY: Cornell University Press, 1990)

Stearn, Ilse F., 'A war-time invention: Lecture by Mrs Ayrton, July 27th', *Girton Review*, Jubilee supplement (1920), 42–43

Stephen, Barbara, *Emily Davies and Girton College* (London: Constable, 1927)

Stopes, Marie ('From a correspondent') 'Women and science: Complaint against the Royal Society: The handicap of sex (from a correspondent)', *The Times*, June 16, 1914, News, p. 5

Sutton, Graham, 'The centenary of the birth of W.H. Young', *Mathematical Gazette*, 59 (1963), 17–21

Swanwick, H., 'Memoir of Girton, 1882–1885', in *Strong-minded Women and Other Lost Voices from Nineteenth-century England*, ed. by Janet Murray (Harmondsworth: Penguin, 1982), pp. 239–242

Tabor, Margaret E., 'Philippa Garrett Fawcett, 1887–1902', *Newnham College Roll Letter* (January 1949), 46–51

Tanner, J.R., ed., *The Historical Register of the University of Cambridge, being a Supplement to the Calendar with a Record of University Honours and Distinctions to the Year 1910* (Cambridge: Cambridge University Press, 1917)

Tattersall, James, McMurran, Shawnee and Cartwright, Mary L., 'An interview with Dame Mary L. Cartwright, D.B.E., F.R.S.', *The College Mathematics Journal*, 32 (4) (2001), 242–254

Tattersall, James J. and McMurran, Shawnee L., 'Hertha Ayrton: A persistent experimenter', *Journal of Women's History*, 7 (2) (1995), 86–112

Tickner, Lisa, 'Suffrage campaigns: The political imagery of the British women's suffrage movement', in *The Edwardian Era*, ed. by Jane Beckett and Deborah Cherry (Oxford: Phaidon, 1987), pp. 100–117

Tobies, Renate, 'In spite of male culture: women in mathematics', in Camina, Rachel and Fajstrup, Lisbeth, eds, *European Women in Mathematics: Proceedings of the 9th General Meeting* (New York: Hindaw, 2007), pp. 25–35

The Times, 'At last there seems to be a fair prospect that …', July 3 1905, Editorial/Leader, p. 9; 'Aspects of Eugenics', July 25 1912, News, p. 9; 'Mrs Hertha Ayrton: A distinguished woman scientist', August 23 1923, Obituaries, p. 11

The Times, The Royal Society Tercentenary compiled from a special supplement of *The Times*, July 1960 (London: *The Times*, 1961)

Thomas, R. Hinton, *Nietzsche in German Politics and Society: 1890–1918* (Manchester: Manchester University Press, 1983)

Tosh, John, 'Domesticity and manliness in the middle-class family of Edward White Benson', in *Manful Assertions: Masculinities in Britain since 1800*, ed. by Michael Roper and John Tosh (London: Routledge, 1991), pp. 44–73

Traweek, Sharon, *Beamtimes and Lifetimes: The World of High Energy Physicists* (Cambridge MA: Harvard University Press, 1988)

Trotter, A.P., 'Mrs Ayrton's work on the electric arc', *Nature*, 113 (January 12 1924), 48–49

Two fellows of the Royal Geographical Society, 'The Royal Geographical Society and women', *The Times*, August 6, 1892, Letters to the Editor, p. 7

Vertinsky, Patricia, 'Exercise, physical capability and the eternally-wounded woman in late nineteenth century North America', *Journal of Sport History*, 14 (1) (1987), 7–27

Vicinus, Martha, ed., *A Widening Sphere: Changing Roles of Victorian Women* (Bloomington: Indiana University Press, 1977)

Vicinus, Martha, *Independent Women: Work and Community for Single Women, 1850–1920* (London: Virago, 1994)

Vickery, Margaret Burney, *Buildings for Bluestockings: The Architecture and Social History of Women's Colleges in late-Victorian England* (Newark: University of Delaware Press, 1999)

Warner, Marina, *Monuments and Maidens: The Allegory of the Female Form* (London: Weidenfeld and Nicolson, 1985)

Warwick, Andrew, 'Exercising the student body: Mathematics and athleticism in Victorian Cambridge', in *Science Incarnate: Historical Embodiments of Natural Knowledge*, ed. by Christopher Lawrence and Steven Shapin (Chicago: University of Chicago Press, 1998), pp. 288–326

Warwick, Andrew, *Masters of Theory: Cambridge and the Rise of Mathematical Physics* (Chicago: University of Chicago Press, 2003)

Weindling, Paul, *Health, Race and German Politics Between National Unification and Nazism, 1870–1945* (Cambridge: Cambridge University Press, 1989)

Weisz, George, *The Emergence of Modern Universities in France, 1863–1914* (Princeton NJ: Princeton University Press, 1983)

Wells, H.G., *Ann Veronica* (London: Everyman/J.M. Dent, 1943; repr. 1999)

Wertheim, Margaret, *Pythagoras' Trousers: God, Physics and the Gender Wars* (London: Fourth Estate, 1997)

Whitman, B.S., 'Ada Isabel Maddison (1869–1950)', in *Women of Mathematics: A Biobibliographic Sourcebook, with a forward by Alice Schafer*, ed. by L.S. Grinstein and P.J. Campbell (Connecticut: Greenwood Press, 1987), pp. 144–146

Wiegand, Sylvia, 'Grace Chisholm Young and William Henry Young: A partnership of itinerant British mathematicians', in *Creative Couples in the Sciences*, ed. by Helena M. Pycior, Nancy G. Slack and Pnina G. Abir-Am (New Brunswick, NJ: Rutgers University Press, 1996), pp. 126–140

Wiegand, S.M., 'Grace Chisholm Young', in *Women in Mathematics*, ed. by L.M. Osen (Cambridge, MA: MIT Press, 1990), pp. 247–253

Wiener, Martin J., *English Culture and the Decline of the Industrial Spirit, 1850–1980* (Cambridge: Cambridge University Press, 1981)

Wilson, David B., 'Experimentalists among the mathematicians: Physics in the Cambridge Natural Sciences tripos, 1851–1900', *Historic Studies in Physical Science*, 12 (2) (1982), 325–371

Wilson, Robin J., 'Hardy and Littlewood', in *Cambridge Scientific Minds*, ed. by Peter Harmant and Simon Mitton (Cambridge: Cambridge University Press, 2002), pp. 202–219

Winter, Alison, 'A calculus of suffering: Ada Lovelace and the bodily constraints on women's knowledge in early Victorian England', in *Science Incarnate: Historical Embodiments of Natural Knowledge*, ed. by Christopher Lawrence and Steven Shapin (Chicago: University of Chicago Press, 1998), pp. 202–239

Woolley, Benjamin, *The Bride of Science: Romance, Reason and Byron's Daughter* (London: Macmillan, 1999)

Wood, Alexander, *The Cavendish Laboratory* (Cambridge: Cambridge University Press, 1946)

Wosk, Julie, *Women and the Machine: Representations from the Spinning Wheel to the Electronic Age* (Baltimore: Johns Hopkins Press, 2001)

Wright, Sir Almroth, 'Militant hysteria', *The Times*, March 28 1912, Letters to Editor

Wynne, David, 'Natural knowledge and social context: Cambridge physicists and the luminous ether', in *Science in Context: Readings in the Sociology of Science*, ed. by Barry Barnes and David Edge (Milton Keynes: Open University Press, 1982), pp. 212–231

Yeo, Richard R., 'Scientific method and the rhetoric of science in Britain, 1830–1917', in *The Politics and Rhetoric of Scientific Method*, ed. by John A. Schuster and Richard R. Yeo (Dordrecht: Reidel, 1986), pp. 259–297

Young, Laurence Chisholm, *Mathematicians and Their Times* (Amsterdam: North-Holland, 1981)

Young, W.H., 'Reply', *Messenger of Mathematics*, 42 (1913), 113

Young, W.H. and Young, Grace Chisholm, *The Theory of Sets of Points* (Cambridge: Cambridge University Press, 1906)

Zangwill, Edith Ayrton, *The Call* (London: Allen and Unwin, 1924)

Zangwill, Israel, 'Professor Ayrton', *The Times*, November 11 1908, Letters to Editor, p. 15

Zeeman, P., 'Scientific worthies: Sir William Crookes, F.R.S.', *Nature*, 77 (November 7, 1907), 1–3

Theses

Garriock, Jean Barbara, *Late Victorian and Edwardian Images of Women and Their Education in the Popular Periodical Press with Particular Reference to the Work of L.T. Meade* (unpublished doctoral thesis, University of Liverpool, 1997)

Gould, Paula, *Femininity and Physical Science in Britain, 1870–1914* (unpublished doctoral thesis, University of Cambridge, 1998)

Online sources

Biographies of Women Mathematicians <http://www.agnesscott.edu/Lriddle/women/women.htm> [accessed April 8 2009]

Davis Archive of Female Mathematicians <http://www-history.mcs.st-andrews.ac.uk/history/Davis/info.html> [accessed March 25 2009]

Institution of Electrical Engineers, *Gertrude Entwisle*, <http://www.iee.org/TheIEE/Research/Archives/Histories&Biographies/Entwisle.cfm> [accessed February 5 2005]; *The Arc Lamp* <http://archives.iee.org/about/Arclamps/arclamps.htm> [accessed February 11 2005]

Irvine, A.D., 'Russell's Paradox', *The Stanford Encyclopedia of Philosophy (Summer 2004 Edition)*, ed. by Edward N. Zalta <http://plato.stanford.edu/archives/sum2004/ entries/russell-paradox/> [accessed February 12 2005)

Jacobs, Joseph and Lipkin, Goodman, 'Hartog, Numa Edward', JewishEncyclopedia.com <http://www.jewishencyclopedia.com/> [accessed February 7 2005]

Mackie, Robin and Roberts, Gerrylynn, *Biographic Database of the British Chemical Community*, 1880–1970 (Milton Keynes: Open University, 2004) <http://www5. open.ac.uk/Arts/chemists/> [accessed February 5 2005]

National Physical Laboratory, *Metromnia*, 9 (2000), <http://www.npl.co.uk/npl/ publications/metromnia/> [accessed February 5 2005]

O'Connor, John J. and Robertson, Edmund F., *The MacTutor History of Mathematics Archive* <http://www-history.mcs.st-andrews.ac.uk/history/index.html> [accessed February 5 2005]

Royal Society of London, *Online Biographies: Sackler Archive Resource* <http:// www.royalsoc.ac.uk/page.asp?id=1724> [accessed February 5 2005]

Tattersall, Jim and McMurran, Shawnee, *Women and the Educational Times* <http://www.math.csusb.edu/faculty/mcmurran/hpm.doc> [accessed February 5 2005]

Thomson and Gale, *The Times Digital Archive, 1785–1985* <http://web2.info-trac.galegroup.com/menu> [accessed February 5 2005]

Tobies, Renate, 'Why a Felix Klein prize?', *ECMI Newsletter*, 27 (March 2000) <http://www.mafy.lut.fi/EcmiNL/> [accessed March 24 2009]

Index

Abbot, Edwin, 20, 189
Abney, William, 186
Aitken, John, 134
alchemy, 190
Allen, Florence E., 166
Althoff, Friedrich, 41
American Mathematical Society, 164–5
Anderson, Dr Elizabeth Garrett, 28, 185
Arber, Agnes, 169
Armstrong, Henry, 80, 81, 89–90, 129, 193, 200
astronomy, 154–5, 157
ayrton fan, 195–6
Ayrton, Hertha (*née* Marks), 2, 5, 10–13, 89–90, 90–1, 167
  and the British Association for the Advancement of Science (BAAS), 198
  at the Central Institution, 82–3, 126–8, 130
  collaboration, 73–6, 81–2
  and eugenics, 79
  at Finsbury Technical College, 71–3
  at Girton College, 14, 15–16, 22–3, 24, 25, 28, 29, 35, 154
  and Hughes Medal, 125–6, 183, 186, 192, 201–2
  and Institution of Electrical Engineers (IEE), 12, 198–9, 202
  Jewish heritage, 11–13, 86, 139, 196–7
  in the laboratory, 127–8, 133–7
  nomination for Royal Society fellowship, 185–8, 199–200
  patents, 15–16, 71, 86, 92, 187
  and Physical Society, 198
  and Royal Society, 70, 75, 76, 81, 91, 110, 133–4, 135, 175, 176–87, 193–5

  and women's suffrage, 131, 139, 176, 191–2, 194, 196–7, 201
Ayrton, William Edward, 12, 13, 76, 79, 81, 121, 207
  at the Central Institution, 77–8
  commercial interests, 80, 128
  marriage to Hertha, 73–5, 82, 89–90, 196
  and women's suffrage, 75

Babbage, Charles, 30, 63
Balfour, Arthur, 79, 140
Barnum, Charlotte C., 168
Barwell, Mildred Emily, 158–62
Bate, Dorothea, 183, 184
Bateson, William, 152, 165, 200
Bedford College, London, 17, 33, 119, 153, 155, 162
Bell, Julia, 155
Bennett, Geoffrey T., 17, 21, 27, 30, 38
Berlin, University of, 40, 42
Berry, Arthur, 22, 37, 98, 101, 102
Birkhoff, George, 100
Blind, Ottilie, 13, 207
Bodichon, Barbara Leigh Smith, 11, 12, 27, 72
Bosworth, Ann Lucy, 41
British Association for the Advancement of Science (BAAS), 197–8
British Astronomical Association (BAA), 198
Browning, Elizabeth Barrett, 112, 113
Bryant, Sophie, 33, 156, 158, 162
Bryn Mawr College, 27, 38, 39, 43, 51, 146
Burstall, Sara A., 10, 16, 18, 21, 22, 30, 115, 153, 210
Butler, Henry Montague, 99

Printed in the United States
By Bookmasters